Trends in Mathematics

Research Perspectives

Research Perspectives collects core ideas and developments discussed at conferences and workshops in mathematics, as well as their increasingly important applications to other fields. This subseries' rapid publication of extended abstracts, open problems and results of discussions ensures that readers are at the forefront of current research developments. Proposals for volumes can be submitted using the online book project submission form at our website www.birkhauser-science.com.

Microlocal Methods in Mathematical Physics and Global Analysis

Daniel Grieser
Stefan Teufel
Andras Vasy

Editors

 Birkhäuser

Editors

Daniel Grieser
Department of Mathematics
University of Oldenburg
Germany

Stefan Teufel
Department of Mathematics
University of Tübingen
Germany

Andras Vasy
Department of Mathematics
Stanford University
Stanford, CA, USA

ISBN 978-3-0348-0465-3   ISBN 978-3-0348-0466-0 (eBook)
DOI 10.1007/978-3-0348-0466-0
Springer Basel Heidelberg New York Dordrecht London

Library of Congress Control Number: 2012951681

Mathematics Subject Classification (2010): 35-XX, 58-XX

Printed on acid-free paper

Springer Basel is part of Springer Science+Business Media (www.springer.com)

Preface

Since its invention in the mid-twentieth century the field of microlocal analysis was characterized by a tight interplay of mathematics and physics. While in language and rigor it is a field of mathematics, many of its ideas originate in physics: Its playing field is the cotangent bundle of a manifold, the mathematical counterpart of the phase space of classical mechanics; its motivating problems came from partial differential equations, particularly those arising in physics, like the Laplace equation and the associated spectral problem, and the wave equation. Among its fundamental tools is the WKBJ method, which was invented for quantum mechanics and also used in geometrical optics; associating an operator to a symbol is one manifestation of quantization. These ideas, often used rather informally in the physics context, were made mathematically precise in microlocal analysis, and often triggered further developments in mathematics, foremost in the field of partial differential equations, but also in symplectic geometry or in singularity theory, to name a few; they were also used in index theory as one of their deepest applications.

In spite of this common ancestry of microlocal analysis in mathematics and physics there are few conferences which bridge the gap between the various communities that use this set of techniques. This is how the idea for the conference 'Microlocal methods in mathematical physics and global analysis' arose. Our aim was to bring together researchers at the highest international level in these areas, in order to foster interaction, inform about new developments and get a state of the art picture.

The major themes of the conference were the use or development of microlocal techniques in semiclassical and adiabatic limits, singular spaces, spectral and scattering theory, wave propagation and topological applications, and we have organized this collection of extended abstracts accordingly, although not every talk admits a unique assignment to one of these categories. A recurrent overall theme of many of the talks was the occurence of singular settings, that is, where the underlying space is singular or non-compact, or where one studies a family of operators or spaces approaching a singular limit. The systematic exploration of such singular problems has been a focus of much research in microlocal analysis since the 1980s.

The present volume is a collection of extended abstracts of most talks at the conference. The talks were given by the first-mentioned authors of each extended abstract. We believe that the format of extended abstract is a valuable means for quick communication of current research, since it allows authors to disseminate their results at an early stage and in a condensed form, and puts an emphasis on conveying the essence of the lectures, without being overburdened by technicalities. Therefore, we hope that this format, which was introduced by the Mathematisches Forschungsinstitut Oberwolfach in 2004 under the name Oberwolfach Reports, will be used more frequently at other conferences as well. We are very grateful to Birkhäuser Verlag for engaging in this new format and agreeing to publish this volume.

The conference could not have happened without the support from various sources: Major funding was supplied by the Deutsche Forschungsgemeinschaft (SFB/Transregio 71, "Geometric PDEs") and the National Science Foundation of the U.S.A. (grant No. 1067924); we are also grateful to the University of Tübingen for hosting the conference.

The organizers
D. Grieser
S. Teufel
A. Vasy

Contents

Part I
Semiclassical and Adiabatic Limits

Local Smoothing with a Prescribed Loss for the Schrödinger Equation

Hans Christianson and Jared Wunsch

1 Introduction: What Is Local Smoothing

In $\mathbb{R}^n$, the Schrödinger propagator at time t is unitary on H^s spaces. However, solutions to the linear Schrödinger equation on $\mathbb{R}^n$ are smoother *on average* in time, and *locally* in space:

$$\int_0^T \left\| \langle x \rangle^{-1/2-\epsilon} e^{it\Delta} u_0 \right\|_{H^{1/2}}^2 dt \leq C \|u_0\|_{L^2}^2.$$

What about different geometries? In [5], Doi showed the $H^{1/2}$ smoothing effect holds on asymptotically Euclidean manifolds if and only if the manifold is non-trapping (meaning all geodesics escape to infinity).

2 Trapping: Hyperbolic Orbits – Local Smoothing

What about trapping geometries? There are several previous results in this direction.

Theorem 1 ([3]). *Let* (M, g) *non-compact, Riemannian manifold with or without a compact boundary, and assume* M *is asymptotically Euclidean (multiple copies is okay) outside a compact set, and assume* $\gamma \subset M$ *is a hyperbolic closed geodesic with only transversal reflections, and* M *non-trapping otherwise. Let* $r = \operatorname{dist}_g(x, x_0)$ *be a "radial" variable. Then for all* $\epsilon > 0$, *there exists* $C > 0$ *such that*

$$\int_0^T \left\| \langle r \rangle^{-1/2-\epsilon} e^{it\Delta_g} u_0 \right\|_{H^{1/2-\epsilon}}^2 dt \leq C \|u_0\|_{L^2}^2.$$

H. Christianson (✉) · J. Wunsch
Department of Mathematics, UNC-Chapel Hill, CB#3250 Phillips Hall, Chapel Hill, NC 27599, USA
e-mail: hans@math.unc.edu; jwunsch@math.northwestern.edu

D. Grieser et al. (eds.), *Microlocal Methods in Mathematical Physics and Global Analysis*, Trends in Mathematics, DOI 10.1007/978-3-0348-0466-0_1, © Springer Basel 2013

This generalizes results of Burq [1] (Ikawa's example), and has been generalized by the first author [2] and also by Datchev [4]. Apart from exhibiting a deep connection between geometry and dispersion, there are additional applications to Strichartz estimates and nonlinear Schrödinger equations.

3   Limitations of Local Smoothing

The theorem of Doi shows if there is no trapping there is a gain of $1/2$ derivative, and if there is trapping you must lose *something* but does not show what you lose. One hyperbolic trapped orbit or a very "thin" trapped set loses a "trivial" ϵ. A stable or *elliptic* geodesic loses everything (no smoothing effect).

An important question is to ask if there is something in between trivial loss and total loss.

We now introduce a class of asymptotically Euclidean examples with a *degenerate* hyperbolic orbit. We consider the manifold $X = \mathbb{R}_x \times \mathbb{R}_\theta / 2\pi\mathbb{Z}$, equipped with a metric of the form

$$g = dx^2 + A^2(x)d\theta^2,$$

where $A \in \mathcal{C}^\infty$ is a smooth function, $A \geq \epsilon > 0$ for some epsilon. We are primarily interested in the case $A(x) = (1 + x^{2m})^{1/2m}$, $m \in \mathbb{Z}_+$, in which case the manifold is asymptotically Euclidean (with two ends). Clairaut's theorem implies the only periodic geodesic is at $x = 0$. $A(x)^{-2}$ has a critical point of order x^{2m} at $x = 0$, which is *degenerate* for $m > 1$. The Gaussian curvature is nonpositive, asymptotically 0 as $x \to \pm\infty$, and vanishes to order $2m - 2$ at $x = 0$.

We prove the following theorem.

Theorem 2 (Local Smoothing with loss). *Suppose X is as above for $m \geq 2$, and assume u solves*

$$\begin{cases} (D_t - \Delta)u = 0 \text{ in } \mathbb{R} \times X, \\ u|_{t=0} = u_0 \in H^s \end{cases}$$

for some $s \geq m/(m + 1)$. Then for any $T < \infty$, there exists a constant $C > 0$ such that

$$\int_0^T \| \langle x \rangle^{-3/2} u\|_{H^1(X)}^2 dt \leq C(\| \langle D_\theta \rangle^{m/(m+1)} u_0\|_{L^2}^2 + \| \langle D_x \rangle^{1/2} u_0\|_{L^2}^2).$$

This theorem says that there is a $1/(m + 1)$ derivative gain in the tangential direction, and $1/2$ derivative gain in the transversal direction.

Theorem 3. *Theorem 2 is sharp, and the estimate can be saturated on a weak semiclassical time scale.*

4   Sketch of Proof Ideas

We use a positive commutator idea: let $B = \arctan(x)\partial_x$ and compute

$$[\Delta, B] = 2\langle x\rangle^{-2}\partial_x^2 + 2A'A^{-3}\arctan(x)\partial_\theta^2 + \text{l.o.t.}$$

Here "l.o.t." can be absorbed into $H^{1/2}$ energy. Everything looks good except the coefficient of ∂_θ^2 has $A'\arctan(x)$, which vanishes to order $2m$ at $x = 0$, so the commutator is not strictly positive! Integrations by parts yields

$$\int_0^T (\|\langle x\rangle^{-1}\partial_x u\|_{L^2}^2 + \||x|^m\langle x\rangle^{-m-3/2}\partial_\theta u\|_{L^2}^2)dt \leq C\|u_0\|_{H^{1/2}}^2. \qquad (1)$$

In order to estimate near $x = 0$, we separate variables:

$$u(t, x, \theta) = \sum_k e^{ik\theta} u_k(t, x),$$

and

$$u_0(x, \theta) = \sum_k e^{ik\theta} u_{0,k}(x)$$

and try to estimate on each mode u_k. By orthogonality, it suffices to show

$$\int_0^T \|\chi(x)k u_k\|_{L^2(\mathbb{R})}^2 dt \leq C(\|\langle k\rangle^{m/(m+1)} u_{0,k}\|_{L^2}^2 + \|u_{0,k}\|_{H^{1/2}}^2$$

for some $\chi \in C_c^\infty(\mathbb{R})$ with $\chi(x) \equiv 1$ near $x = 0$.

By a duality argument, energy cutoff, and Fourier transform $t \mapsto \tau$, it suffices to show the following (sharp) cutoff resolvent estimate.

Proposition 1. *With the notations above, the operator Q_k satisfies the estimate*

$$\|\phi(x, D/|k|)(\tau + Q_k)^{-1}\phi(x, D/|k|)\|_{L_x^2 \to L_x^2} \leq Ck^{-2/(m+1)}$$

where $\phi \in C_c^\infty(T^*\mathbb{R})$ with $\phi \equiv 1$ near $(0, 0)$.

The proof of this proposition involves careful analysis of escape function dynamics and a sophisticated two-parameter semiclassical calculus to estimate operators at the level of the uncertainty principle.

References

1. N. Burq. Smoothing effect for Schrödinger boundary value problems. *Duke Math. J.*, 123(2):403–427, 2004.
2. Hans Christianson. Cutoff resolvent estimates and the semilinear Schrödinger equation. *Proc. Amer. Math. Soc.*, 136:3513–3520, 2008.

3. Hans Christianson. Dispersive estimates for manifolds with one trapped orbit. *Comm. Partial Differential Equations*, 33:1147–1174, 2008.
4. Kiril Datchev. Local smoothing for scattering manifolds with hyperbolic trapped sets. *Comm. Math. Phys.*, 286(3):837–850, 2009.
5. Shin-ichi Doi. Smoothing effects of Schrödinger evolution groups on Riemannian manifolds. *Duke Math. J.*, 82(3):679–706, 1996.

Propagation Through Trapped Sets and Semiclassical Resolvent Estimates

Kiril Datchev and András Vasy

Let $P = -h^2\Delta + V(x)$, $V \in C_0^\infty(\mathbb{R}^n)$. We are interested in semiclassical resolvent estimates of the form

$$\|\chi(P - E - i0)^{-1}\chi\|_{L^2(\mathbb{R}^n) \to L^2(\mathbb{R}^n)} \leq \frac{a(h)}{h}, \qquad h \in (0, h_0], \tag{1}$$

for $E > 0$, $\chi \in C^\infty(\mathbb{R}^n)$ with $|\chi(x)| \leq \langle x \rangle^{-s}$, $s > 1/2$. We ask: how is the function $a(h)$ for which (1) holds affected by the relationship between the support of χ and the trapped set at energy E, defined by

$$K_E = \{\alpha \in T^*\mathbb{R}^n : \exists C > 0, \forall t > 0, |\exp(tH_p)\alpha| \leq C\}?$$

Here $p = |\xi|^2 + V(x)$ and $H_p = 2\xi \cdot \nabla_x - \nabla V \cdot \nabla_\xi$.

We have (1) with $\chi(x) = \langle x \rangle^{-s}$ and $a(h) = C$ for all E in a neighborhood of $E_0 > 0$ if and only if $K_{E_0} = \emptyset$ ([6,7]). For general V and χ, the optimal bound is $a(h) = \exp(C/h)$, but Burq [1] and Cardoso-Vodev [2] prove that for any given V, if χ vanishes on a sufficiently large compact set, for any $E > 0$ there exists C such that (1) holds with $a(h) = C$. In our main theorem we improve the condition on χ and obtain a shorter proof at the expense of an a priori assumption.

Theorem 1 ([3]). *Fix $E > 0$. Suppose that (1) holds for $\chi(x) = \langle x \rangle^{-s}$ with $s > 1/2$ and with $a(h) = h^{-N}$ for some $N \in \mathbb{N}$. Then if we take instead χ such that $K_E \cap T^* \operatorname{supp} \chi = \emptyset$, we have (1) with $a(h) = C$.*

K. Datchev (✉)
Mathematics Department, MIT, Cambridge, MA, 02139-4307, USA
e-mail: datchev@math.mit.edu

A. Vasy
Mathematics Department, Stanford University, Stanford, CA, 94305-2125, USA
e-mail: andras@math.stanford.edu

D. Grieser et al. (eds.), *Microlocal Methods in Mathematical Physics and Global Analysis*,
Trends in Mathematics, DOI 10.1007/978-3-0348-0466-0_2, © Springer Basel 2013

In fact our result holds for more general operators, and the cutoff χ can be replaced by a cutoff in phase space whose microsupport is disjoint from K_E. In certain situations it is even possible to take a cutoff whose support overlaps K_E: see [3] for more details and references.

The a priori assumption that (1) holds for $\chi(x) = \langle x \rangle^{-s}$ with $a(h) = h^{-N}$ is not present in [1, 2] and is not always satisfied, but there are many examples of hyperbolic trapping where it holds: see e.g. [5, 8].

To indicate the comparative simplicity of our method, we prove a special case of the Theorem, under the additional assumption that $\operatorname{supp} V \subset \{|x| < R_0\}$ and $\operatorname{supp} \chi \subset \{R_0 < |x| < R_0 + 1\}$. In other words, suppose $(P - \lambda)u = f$, with $\operatorname{Re} \lambda = E$, and $\operatorname{supp} f \subset \{R_0 < |x| < R_0 + 1\}$, $\|f\| \leq 1$. We must prove that $\|\chi u\| \leq Ch^{-1}$, uniformly as $\operatorname{Im} \lambda \to 0^+$. Here and below all norms are L^2 norms.

Let S denote functions in $C^\infty(T^*\mathbb{R}^n)$ which are bounded together with all derivatives, and for $a \in S$ define

$$\operatorname{Op}(a)u(x) = (2\pi h)^{-n} \int \exp(i(x - y) \cdot \xi/h)a(x, \xi)u(y)dyd\xi.$$

Because $P - \lambda$ has a semiclassical elliptic inverse away from $p^{-1}(E)$ (see for example [4, Chap. 4]), we have $\|\operatorname{Op}(a)u\| \leq C$ whenever $\operatorname{supp} a \cap p^{-1}(E) = \emptyset$. Consequently it is enough to show that $\|\operatorname{Op}(a)u\| \leq Ch^{-1}$ for some $a \in S$ with a nowhere vanishing on $T^* \operatorname{supp} \chi \cap p^{-1}(E)$. We will prove this inductively: we will show that if there is a_1 with this nowhere vanishing property such that $\|\operatorname{Op}(a_1)u\| \leq Ch^k$, then there is a_2 with the same nowhere vanishing property such that $\|\operatorname{Op}(a_2)u\| \leq Ch^{k+1/2}$, provided $k \leq -3/2$. The base case follows from the a priori assumption that $\|u\| \leq h^{-N-1}$, so it suffices to prove the inductive step.

Take $\varphi = \varphi(|x|) \geq 0$ a smooth function such that $\varphi = 1$ when $|x| \leq R_0$, $\varphi = 0$ when $|x| \geq R_0 + 1$, $\varphi' = -\psi^2$ with ψ smooth. We require further that $T^* \operatorname{supp} \psi$ be contained in the set where a_1 is nonvanishing, and in the end we will take $a_2 = \psi$. We will now use a positive commutator argument with φ as the commutant:

$$i\langle [P, \varphi]u, u \rangle = i\langle u, \varphi f \rangle - i\langle \varphi f, u \rangle - 2\operatorname{Im} \lambda \|u\|^2 \geq -C\|\psi u\|\|f\|, \quad (2)$$

where we used first $(P - \lambda)u = f$ and then $\operatorname{Im} \lambda \geq 0$ and $\operatorname{supp} f \subset \{\psi \neq 0\}$. The semiclassical principal symbol of $i[P, \varphi]$ is

$$hH_p\varphi = 2h\rho\varphi' = -2h\rho\psi^2,$$

where ρ is the dual variable to $|x|$ in $T^*\mathbb{R}^n$.

We now define an open cover and partition of unity of $T^* \operatorname{supp} \chi$ according to the regions where this commutator does and does not have a favorable sign (the favorable sign is $H_p\varphi < 0$, because of the direction of the inequality in (2)). Take $c > 0$ small enough that for $\rho < 2c$, $|x| > R_0$, $t < 0$ we have $x + 2\rho t \notin \operatorname{supp} V$. Let K be a neighborhood of $p^{-1}(E) \cap T^* \operatorname{supp} \chi$ with compact closure in $T^*\{R_0 < |x| < R_0 + 1\}$, and let O be a neighborhood of K with compact closure in $T^*\{R_0 < |x| < R_0 + 1\}$, and let

$$U_+ = \{\alpha \in O : \rho > c\}, \qquad U_- = \{\alpha \in O : \rho < 2c\} \cup (T^*\mathbb{R}^n \setminus K).$$

Take $\phi_\pm \in C_0^\infty(O)$ with $\phi_+^2 + \phi_-^2 = 1$ on $T^* \operatorname{supp} \chi$ and with $\operatorname{supp} \phi_\pm \subset U_\pm$. Then

$$H_p \varphi = -b^2 - 2\rho\psi^2\phi_-^2, \qquad \text{where } b = \sqrt{2\rho}\psi\phi_+,$$

and if $B = \operatorname{Op}(b)$ and $\Phi_- = \operatorname{Op}(\phi_-)$

$$i[P, \varphi] = -hB^*B + h\Phi_- R_1 \Phi_- + h^2 R_2 + O(h^\infty),$$

where $R_{1,2} = \operatorname{Op}(r_{1,2})$ for $r_{1,2} \in S$ with $\operatorname{supp} r_{1,2} \subset \operatorname{supp} \psi$. Combining with (2), and using L^2 boundedness of R_1, we obtain

$$h\|Bu\|^2 \le Ch\|\Phi_- u\|^2 + h^2\langle R_2 u, u\rangle + C\|\psi u\|\|f\| + O(h^\infty).$$

Since $\langle R_2 u, u\rangle \le Ch^{2k}$ by inductive hypothesis, we have

$$\|Bu\|^2 \le C(\|\Phi_- u\|^2 + h^{2k+1} + h^{-1}\|\psi u\|\|f\|)$$
$$\le C(\|\Phi_- u\|^2 + h^{2k+1} + \delta^{-1}h^{-2} + \delta\|\psi u\|^2),$$

where we used $\|f\| \le 1$, and where $\delta > 0$ will be specified presently. Since at least one of B and Φ_- is elliptic at each point in the interior of $T^* \operatorname{supp} \psi$, we have

$$\|\psi u\|^2 \le C(\|\Phi_- u\|^2 + \|Bu\|^2), \tag{3}$$

from which we conclude that, if δ is sufficiently small,

$$\|Bu\|^2 \le C_\delta(\|\Phi_- u\|^2 + h^{-2} + h^{2k+1}). \tag{4}$$

Because c was chosen small enough that all backward bicharacteristics through $\operatorname{supp} \phi_-$ stay in $T^*\{|x| > R_0\}$, where $P = -h^2\Delta$, we have

$$\|\Phi_- u\| \le Ch^{-1},$$

by standard nontrapping estimates (see, for example, [3, Sect. 6]). This, combined with (3) and (4), gives

$$\|\psi u\|^2 \le C_\delta(h^{-2} + h^{2k+1}),$$

after which taking $a_2 = \psi$ completes the proof of the inductive step.

Acknowledgements The first author is partially supported by a National Science Foundation postdoctoral fellowship, and the second author is partially supported by the NSF under grant DMS-0801226, and a Chambers Fellowship from Stanford University. The authors are grateful for the hospitality of the Mathematical Sciences Research Institute, where part of this research was carried out.

References

1. Nicolas Burq. Lower bounds for shape resonances widths of long range Schrödinger operators. *Amer. J. Math.*, 124:4, 677–735, 2002.
2. Fernando Cardoso and Georgi Vodev. Uniform estimates of the resolvent of the Laplace-Beltrami operator on infinite volume Riemannian manifolds II. *Ann. Henri Poincaré*, 3:4, 673–691, 2002.
3. Kiril Datchev and András Vasy. Propagation through trapped sets and semiclassical resolvent estimates. To appear in *Ann. Inst. Fourier*. Preprint available at arXiv:1010.2190.
4. Lawrence C. Evans and Maciej Zworski. Lecture notes on semiclassical analysis. Available online at http://math.berkeley.edu/~zworski/semiclassical.pdf.
5. Stéphane Nonnenmacher and Maciej Zworski. Quantum decay rates in chaotic scattering. *Acta Math.* 203:2, 149–233, 2009.
6. Didier Robert and Hideo Tamura. Semiclassical estimates for resolvents and asymptotics for total scattering cross-sections. *Ann. Inst. H. Poincaré Phys. Théor.* 46:4, 415–442, 1987.
7. Xue Ping Wang. Semiclassical resolvent estimates for N-body Schrödinger operators. *J. Funct. Anal.* 97:2, 466–483, 1991.
8. Jared Wunsch and Maciej Zworski. Resolvent estimates for normally hyperbolic trapped sets. To appear in *Ann. Inst. Henri Poincaré (A)*. Preprint available at arXiv:1003.4640.

Space–Adiabatic Theory for Random–Landau Hamiltonian: Results and Prospects

Giuseppe De Nittis

The study of the (integer) *Quantum Hall Effect* (QHE) requires a careful analysis of the spectral properties of the $2D$, single-electron Hamiltonian

$$H_{\Gamma,B} := \left(-\mathrm{i}\partial_x - B\,y\right)^2 + \left(-\mathrm{i}\partial_y + B\,x\right)^2 + V_\Gamma(x,y) \tag{1}$$

where $H_B := H_{\Gamma,B} - V_\Gamma$ is the usual *Landau Hamiltonian* (in symmetric gauge) with *magnetic field* B and V_Γ is a $\Gamma \equiv \mathbb{Z}^2$ *periodic potential* which models the electronic interaction with a crystalline structure. Under usual conditions (e.g., $V_\Gamma \in L^2_{\mathrm{loc}}(\mathbb{R}^2)$) the Hamiltonian (1) is self-adjoint on a suitable domain of $L^2(\mathbb{R}^2)$. A direct analysis of the fine spectral properties of (1) is extremely difficult and one needs resorting to simpler effective models hoping to capture (some of) the main physical features in suitable physical regimes.

Weak magnetic field limit. The regime $B \ll 1$ is very interesting since it is easily accessible to experiments. The common lore, (cf. works of R. Peierls, P. G. Harper and D. Hofstadter), says that the "local description" of the spectrum of (1) is "well approximated" by the spectrum of the (effective) *Hofstadter model*

$$\left(H^{(B)}_{\mathrm{Hof}}\xi\right)_{n,m} := e^m_B\,\xi_{n+1,m} + \overline{e^m_B}\,\xi_{n-1,m} + \overline{e^n_B}\,\xi_{n,m+1} + e^n_B\,\xi_{n,m-1} \tag{2}$$

with $\{\xi_{n,m}\} \in \ell^2(\mathbb{Z}^2)$ and $e^m_B := \mathrm{e}^{\mathrm{i}2\pi m B}$.

The above discussion leads to the following questions:

(Q.1) In what mathematical sense are $\mathrm{Spec}(H_{\Gamma,B})$ and $\mathrm{Spec}(H^{(B)}_{\mathrm{Hof}})$ "locally equivalent"?

(Q.2) What is the relation between the "effective" dynamics induced by $H^{(B)}_{\mathrm{Hof}}$ and the "true" dynamics induced by $H_{\Gamma,B}$?

G. De Nittis (✉)
LAGA - Université Paris 13 - Institut Galilée, 93430, Villetaneuse, France
e-mail: denittis@math.univ-paris13.fr

D. Grieser et al. (eds.), *Microlocal Methods in Mathematical Physics and Global Analysis*, Trends in Mathematics, DOI 10.1007/978-3-0348-0466-0_3, © Springer Basel 2013

A third question concerns the rôle of the disorder in the explanation of the QHE. Indeed, the introduction of a *random potential* V_ω (e.g., an *Anderson potential*) in (1), leading to

$$H_{\Gamma,B,\omega} := H_{\Gamma,B} + V_\omega, \tag{3}$$

is essential in order to explain the emergence of the quantum Hall plateaus. Then:

(Q.3) In presence of disorder is it still possible to derive a "simplified" (i.e., effective) model for $H_{\Gamma,B,\omega}$ which encodes the (main) spectral and dynamical properties of the original model?

In order to answer questions (Q.1) and (Q.2) one needs to prove the so-called *Peierls substitution*. This is an old problem which dates back to the works of Bellissard [1] and Helffer and Sjöstrand [6]. However, these works provide only a partial answer to (Q.1) (*local isospectrality*) and no answer for (Q.2). A complete solution has been given only recently by the author and M. Lein in [3]. In this paper a strong version of the Peierls substitution has been derived by means of a joint application of the *Space-adiabatic perturbation theory* (SAPT) developed by Panati et al. [8] and the *magnetic Weyl quantization* developed by Măntoiu and Purice [7]. The main result derived in [3] can be stated as follows:

Theorem 1. *Assume the existence of a $S \subset \mathrm{Spec}(H_{\Gamma,B=0})$ separated from the rest of the spectrum $\mathrm{Spec}(H_{\Gamma,B=0}) \setminus S$ by gaps.*[1] *Then:*

(i) *Associated to S there exists an an orthogonal projection Π_B in $L^2(\mathbb{R}^2)$ such that for any $N \in \mathbb{N}$*

$$\left\| [H_{\Gamma,B}; \Pi_B] \right\| \leqslant C_N \, B^N \qquad if \qquad B \to 0 \tag{4}$$

where $C_N > 0$ are suitable constants. The space $\mathrm{Ran}\,\Pi_B \subset L^2(\mathbb{R}^2)$ is called almost-invariant *subspace.*

(ii) *There exists a* reference *Hilbert space $\mathcal{H}_\mathrm{r}$ (B-independent), an effective (bounded) operator H_B^{eff} on $\mathcal{H}_\mathrm{r}$ and a unitary operator $U_B : \mathrm{Ran}\,\Pi_B \to \mathcal{H}_\mathrm{r}$ such that for any $N \in \mathbb{N}$*

$$\left\| \left(\mathrm{e}^{itH_{\Gamma,B}} - U_B^{-1} \, \mathrm{e}^{itH_B^{\mathrm{eff}}} \, U_B \right) \Pi_B \right\| \leqslant C_N \, B^N \, |t| \qquad if \qquad B \to 0. \tag{5}$$

(iii) *If S corresponds to a single Bloch energy band E_* for the periodic operator $H_{\Gamma,B=0}$, then $\mathcal{H}_\mathrm{r} \equiv \ell^2(\mathbb{Z}^2)$. Moreover if the dispersion law for E_* can be approximated as $E_*(k_1,k_2) = 2\cos(k_1) + 2\cos(k_2) + Bf(k_1,k_2)$, with k_1 and k_2 the Bloch momenta, then*

$$H_B^{\mathrm{eff}} = H_{\mathrm{Hof}}^{(B)} + \mathcal{O}(B) \qquad if \qquad B \to 0. \tag{6}$$

---

[1]This assumption can be relaxed by introducing the notion of *adiabatically decoupled* energy subspace, cf. [8] or [3].

Theorem 1 implies the following answers for (Q.1) and (Q.2): $\Pi_B H_{\Gamma,B} \Pi_B$ and $H_{\text{Hof}}^{(B)}$ are unitarily equivalent up to an error which goes to zero if $B \to 0$ (*asymptotic unitary equivalence*); the dynamics generated by $H_{\text{Hof}}^{(B)}$ approximates the dynamics generated by $\Pi_B H_{\Gamma,B} \Pi_B$ up to a small error over any macroscopic time-scale $t \in [0, T]$.

Question (Q.3) suggests to combine SAPT-techniques with the randomness induced by V_ω. However, one of the main ingredients of SAPT is the separation in fast and slow degrees of freedom induced by the periodic structure of $H_{\Gamma,B=0}$. This separation (mathematically highlighted by a Bloch-Floquet transform) identifies the fast part of the dynamics with the one inside the fundamental cell of Γ. The slow part is related to the motion at the boundary of adjacent cells and is controlled by the slow variation of the Bloch momenta induced by the weak, but non-zero, magnetic field $B \ll 1$. In order to include V_ω in this schema, one needs to assume that the randomness perturbs the periodic structure on a scale larger that the typical length of the crystal and which becomes larger and larger when $B \to 0$. In other words SAPT-tecniques are compatible only with B-dependent random potentials of type

$$V_{\omega,B}(x,y) := w_\omega \big(B^{-1}x, B^{-1}x\big) \tag{7}$$

where w_ω are suitable random variables. In order to overcome the quite unphysical restriction (7) one has to replace the usual Bloch-Floquet transform with some more general tool able to take care of the loss of the translation symmetry. An hint in this direction is provided by the Bellissard's idea of replacing the Bloch-Floquet decomposition with the algebraic notion of *non-commutative Brillouin zone* (i.e. crossed product C^*-algebra) [2].

Strong magnetic field limit. The opposite regime of a strong magnetic field $B \gg 1$ (accessible to experiments by means of optical lattices) is mathematically easier to treat. In this regime, the dominant terms for the "renormalized" Hamiltonian $B^{-1} H_{\Gamma,B}$ turns out to be a harmonic oscillator which fixes the energy threshold (Landau level). Under the assumption $V_\Gamma(x,y) \simeq 2\cos(x) + 2\cos(y)$, the first relevant term for the asymptotic description of the spectral properties of (1) is given by the (effective) *Harper model*

$$\big(H_{\text{Har}}^{(B)}\xi\big)(s) := \xi\big(s + B^{-1}\big) + \xi\big(s - B^{-1}\big) + 2\cos(2\pi s)\xi(s), \qquad \xi \in L^2(\mathbb{R}). \tag{8}$$

The effective operator (8) was firstly derived by M. Wilkinson (1987). However, a rigorous (asymptotic) unitary derivation of $H_{\text{Har}}^{(B)}$ from $H_{\Gamma,B}$, in the spirit of the Theorem 1, has been established only recently in [4]. Moreover, the proof can be easily extended to the case of a random potential V_ω generalizing a perturbative technique developped in [5].

Acknowledgements The author would like to thank F. Klopp for many stimulating discussions. Project supported by the grant ANR-08-BLAN-0261-01.

References

1. J. V. Bellissard. C*-algebras in solid state physics: 2D electrons in uniform magnetic field. In: *Operator Algebras and Applications Vol. 2: Mathematical Physics and Subfactors*, E. Evans et al. editors, *London Mathematical Society Lecture Note Series vol. 136*, Cambridge University Press, 49–76 (1989).
2. J. V. Bellissard, H. Schulz-Baldes, A. van Elst. The noncommutative geometry of the quantum Hall effect. *J. Math. Phys.* **35**, 5373–5471 (1994).
3. G. De Nittis, M. Lein. Applications of Magnetic Ψ-DO Techniques to Space-adiabatic Perturbation Theory. *Rev. Math. Phys.* **23** (3), 233–260 (2011).
4. G. De Nittis, G. Panati. Effective models for conductance in magnetic fields: derivation of Harper and Hofstadter models. arXiv:1007.4786v1 [math-ph].
5. A. Eckstein. Unitary reduction for the two-dimensional Schrödinger operator with strong magnetic field. *Mathematische Nachrichten* **282**, (2009).
6. B. Helffer, J. Sjöstrand. Équation de Schrödinger avec champ magnétique et équation de Harper. In *Schrödinger Operators (Proceedings of the Nordic Summer School in Mathematics, Sandbjerg Slot, Denmark, 1988)*, Lecture Notes in Physics, vol. 345, Springer, 118–197 (1989).
7. M. Măntoiu, R. Purice. The Magnetic Weyl Calculus *J. Math. Phys.* **45**(4), 233–260 (2004).
8. G. Panati, H. Spohn, S. Teufel. Effective dynamics for Bloch electrons: Peierls substitution and beyond. *Comm. Math. Phys.* **242**, 547–578 (2003).
9. M. Wilkinson, An exact effective hamiltonian for a perturbed Landau level, *J. Phys.* **A 20**, 1761–1771 (1987).

Microlocal Analysis of FIOs with Singularities

Raluca Felea

Overview In this talk we describe the composition calculus of Fourier Integral Operators (FIOs) with fold and cusp singularities. Such operators appear in many inverse scattering problems, where the composition calculus can be used as a tool for recovering images. In these problems, caustics occur and create artifacts which make the reconstruction more complicated and challenging. The goal is to understand these artifacts, find their strength and try to remove them.

Motivation In the seismic problems [13, 14], acoustic waves are generated at the surface of the earth, scatter the heterogeneities of the subsurface and return to the surface. The pressure field is measured at the surface and is used to reconstruct an image of the subsurface. We consider the linearized operator F which maps singular perturbations of a smooth background sound speed in the subsurface, to perturbations of the pressure field. We study different cases where the waves are generated from a fixed single source and from a moving single source. In reality caustics appear. We define a caustic in the following way: a ray departing from a source s in the direction α reaches at time t a point denoted $x(t, \alpha)$ in the subsurface. If the projection map $(t, \alpha) \to x(t, \alpha)$ is singular then $x(t, \alpha)$ lies on a caustic. We consider only fold and cusp caustics which occur when this map exhibits fold or cusp singularities. In order to reconstruct the image we apply the operator F^* to the data. In the case when no caustics occur, F^*F is a pseudodifferential operator which can be inverted [1]. The focus is to understand the reconstruction operator F^*F when caustics are present.

Background Let $I^m(X, Y; C)$ be the class of Fourier integral operators, $F : E'(Y) \to D'(X)$ associated to a canonical relation $C \subset (T^*X \times T^*Y) \setminus \{0\}$. Under certain geometric conditions like transverse and clean intersection conditions, this class is closed under composition [2, 11]. When these conditions fail to be satisfied,

R. Felea (✉)
Rochester Institute of Technology, School of mathematical sciences, Rochester, NY, 14620, USA
e-mail: rxfsma@rit.edu

D. Grieser et al. (eds.), *Microlocal Methods in Mathematical Physics and Global Analysis*, 15
Trends in Mathematics, DOI 10.1007/978-3-0348-0466-0_4, © Springer Basel 2013

the geometry of each canonical relation plays an important role in establishing a composition calculus. Let π_L and π_R be the projections to the left and right from C to $T^*X \setminus 0$ and $T^*Y \setminus 0$.

Singularities My interest is in the case when both projections are singular in specific ways: π_L and π_R have both fold singularities (when C is called a two sided fold) or π_R is a submersion with folds and π_L has a cross cap singularity (when C is called a folded cross cap) or π_R and π_L have both cusp singularities (when C is called a two sided cusp). In the above examples, the operator F^*F is not an FIO anymore but it belongs to a new class of operators associated to a pair of intersecting Lagrangeans (Δ, C) where Δ is the diagonal in $T^*X \times T^*Y$ and C is the artifact. In the first two cases, C is a smooth canonical relation and intersects Δ cleanly and we show that F^*F belongs to $I^{p,l}(\Delta, C)$, a class of distributions studied in [8–10, 12] while in the last case, C is a singular Lagrangean called an open umbrella [5, 7].

Fixed source, fold caustics In the case when a single source generates acoustics waves in the presence of fold caustics only, Nolan showed that F is associated to a two sided fold [13]. We proved that $F^*F \in I^{2m,0}(\Delta, \hat{C})$ where $\hat{C}$ is another two sided fold [3, 13]. Using the properties of $I^{p,l}$ classes, this implies that $F^*F \in I^{2m}(\Delta \setminus \hat{C})$ and $F^*F \in I^{2m}(\hat{C} \setminus \Delta)$. Thus the artifact $\hat{C}$ has the same strength as the pseudodifferential part and it cannot be removed.

Fixed source, cusp caustics In the single source case, in the presence of cusp caustics, F is associated to a two sided cusp and $F^*F \in I^{2m}(\Delta, \Lambda)$ [5], where Λ is an open umbrella. We also obtain that away from $\Delta \cap \Lambda$, F^*F is of order $2m$ on both Δ and Λ which means that the artifact is as strong as the initial location of the singularities.

Moving source, fold caustics We prove that under the fold caustic assumption, F is an FIO associated to a folded cross cap canonical relation C, and that $F^*F \in I^{2m-\frac{1}{2},\frac{1}{2}}(\Delta, \tilde{C})$ where $\tilde{C}$ is a two sided fold [4]. In this case $F^*F \in I^{2m}(\Delta \setminus \tilde{C})$ and $F^*F \in I^{2m-\frac{1}{2}}(\tilde{C} \setminus \Delta)$, hence the artifact is $\frac{1}{2}$ smoother and there is hope for the image recovering. So far we obtained H^s estimates for operators belonging to $I^{p,l}(\Delta, \hat{C})$ and $I^{p,l}(\Delta, \tilde{C})$ [6].

Open problems We would like to find Sobolev estimates for the operators in the class $I^{2m}(\Delta, \Lambda)$ and to invert the operators from $I^{2m-\frac{1}{2},\frac{1}{2}}(\Delta, \tilde{C})$.

References

1. Beylkin, G. Imaging of discontinuities in the inverse problem by inversion of a generalized Radon transform, *Jour. Math. Phys.* 28 (1985), 99–108.
2. Duistermaat, J.J., Guillemin, V., The spectrum of positive elliptic operators and periodic bicharacteristics, *Inv. math.*, 29 (1975), 39–79.
3. Felea, R. Composition calculus of Fourier integral operators with fold and blowdown singularities, *Comm. P.D.E*, 30 (13) (2005), 1717–1740.

4. Felea, R., Greanleaf, A., An FIO calculus for marine seismic imaging: folds and cross caps, *Communications in PDEs*, 33 (1), (2008), 45–77.

5. Felea, R., Greanleaf, A., Fourier integral operators with open umbrellas and seismic inversion for cusp caustics, *Math Ress Lett*, 17 (5) (2010), 867–886.

6. Felea, R., Greenleaf, A., Pramanik, M., An FIO calculus for marine seismic imaging, II: Sobolev estimates, *Math. Annalen*, (2011).

7. Givental, A. Lagrangian imbeddings of surfaces and unfolded Whitney umbrella. (English) *Func. Anal. Appl.* 20 (3) (1986), 197–203.

8. Greenleaf, A., Uhlmann, G., Estimates for singular Radon transforms and pseudodifferential operators with singular symbols, *Jour. Func. Anal.*, 89 (1990), 220–232.

9. Greenleaf, A., Uhlmann,G., Composition of some singular Fourier integral operators and estimates for restricted X-ray transforms, *Ann. Inst. Fourier, Grenoble*, 40 (1990), 443–466.

10. Guillemin, V., Uhlmann, G., Oscillatory integrals with singular symbols. *Duke Math. J.* 48 (1) (1981), 251–267.

11. Hörmander, L., Fourier integral operators, I. *Acta mathematica*, 127 (1971), 79–183.

12. Melrose, R.B., Uhlmann, G.A., Lagrangian intersection and the Cauchy problem. *Comm. Pure Appl. Math.*, 32 (4) (1979), 483–519.

13. Nolan, C.J., Scattering in the presence of fold caustics. *SIAM J. Appl. Math.* 61 (2) (2000), 659–672.

14. Nolan, C.J., Symes, W.W., Global solutions of a linearized inverse problem for the acoustic wave equation, *Comm. in PDE* 22, (1997), 919–952.

A Nonlinear Adiabatic Theorem for Coherent States

Clotilde Fermanian Kammerer and Rémi Carles

1 Introduction

We present a result obtained in collaboration with Rémi Carles on the propagation of coherent states for a 1-d cubic nonlinear Schrödinger equation in a semi-classical regime ($\varepsilon \ll 1$):

$$i\varepsilon\partial_t\psi^\varepsilon + \frac{\varepsilon^2}{2}\partial_x^2\psi^\varepsilon = V(x)\psi^\varepsilon + \varepsilon^\alpha|\psi^\varepsilon|^2\psi^\varepsilon, \quad \psi^\varepsilon : \mathbf{R}_t \times \mathbf{R}_x \to \mathbf{C}^N, \quad \alpha > 0.$$

Typically, when $N = 2$, such systems model a binary mixture of Bose-Einstein condensates (double condensate), under the effect of trap potentials (see [11], and [2] – and the references therein – for numerical analysis of this equation).

The initial data is a semi-classical wave packets (coherent states) of the form

$$\psi^\varepsilon(0, x) = \frac{1}{\varepsilon^{1/4}}a\left(\frac{x - x_0}{\sqrt{\varepsilon}}\right)e^{i\xi_0\cdot(x-x_0)/\varepsilon}\chi(x),$$

where $a \in \mathcal{S}(\mathbf{R})$ and $\chi(x)$ is an eigenvector of $V(x)$, $V(x)\chi(x) = \lambda(x)\chi(x)$. Our aim is to discuss under which conditions the solution $\psi^\varepsilon(t, x)$ is asymptotic to a coherent state of the same form as the data.

In the scalar linear case ($\Lambda = 0$, $V(x) = \lambda(x)\text{Id}$), such asymptotics are valid until the celebrated Ehrenfest time: $T(\varepsilon) \propto \text{Log}\left(\frac{1}{\varepsilon}\right)$ (see [1, 6, 7] and [10]). In the scalar nonlinear case: $V(x) = \lambda(x)\text{Id}$, a similar result holds if $d = 1$, while for

C.F. Kammerer (✉)
Université Paris EST, UMR CNRS 8050, 94010, Créteil Cedex, France
e-mail: Clotilde.Fermanian@univ-paris12.fr

R. Carles
Université Montpellier 2, UMR CNRS 5149, F-34095, Montpellier, France
e-mail: Remi.Carles@math.cnrs.fr

D. Grieser et al. (eds.), *Microlocal Methods in Mathematical Physics and Global Analysis*, 19
Trends in Mathematics, DOI 10.1007/978-3-0348-0466-0_5, © Springer Basel 2013

$d > 1$, the asymptotics hold only on shorter times: $\tilde{T}(\varepsilon) \propto \mathrm{LogLog}\left(\frac{1}{\varepsilon}\right)$ (see [4]). Finally, in the matrix-valued linear case, the result is still valid under a spectral gap condition (see [13, 14] and [12]) for results on adiabatic decoupling and [8, 9] for precise studies of coherent states).

In the nonlinear matrix-valued case, we perform the following assumptions:

(H1) *A weak spectral gap condition*: let $(\lambda_j(x))_{1 \leq j \leq P}$ be the eigenvalues of $V(x)$,

$$\exists c_0, n_0 \in \mathbf{R}^+, \quad \forall j \neq k, \quad |\lambda_j(x) - \lambda_k(x)| \geq c_0 \langle x \rangle^{-n_0}.$$

(H2) *A weak linear coupling*: The potential $V(x)$ is of the form $V(x) = D(x) + W(x)$ with D diagonal with coefficients at most quadratic and W symmetric bounded as well as its derivatives.

Then, we are able to prove that ψ^ε is asymptotic to a coherent state which remains in the same eigenspace of V, for times of order $\mathrm{LogLog}\left(\frac{1}{\varepsilon}\right)$. Besides, under precise conditions on the energy of the wave packets, we can prove a superposition result.

2 Notations

Let us introduce first a few notations. We define the Hamiltonian flow

$$\begin{cases} \dot{x}(t) = \xi(t) & ; \quad x(0) = x_0, \\ \dot{\xi}(t) = -\lambda'(x(t)) & ; \quad \xi(0) = \xi_0. \end{cases}$$

Note that $x, \xi \in C^\infty(\mathbf{R}; \mathbf{R}^d)$ and that the subquadraticity of $V(x)$ yields an exponential control in time on the norm of the trajectories. We define the classical action as

$$S(t) = \int_0^t \left(\frac{1}{2}|\xi(s)|^2 - \lambda(x(s))\right) ds,$$

and we perform the rescaling:

$$\psi^\varepsilon(t, x) \sim \frac{1}{\varepsilon^{1/4}} u^\varepsilon\left(t, \frac{x - x(t)}{\sqrt{\varepsilon}}\right) e^{i(S(t) + \xi(t) \cdot (x - x(t)))/\varepsilon} \chi(t, x).$$

In the scalar case, $V(x) = \lambda(x)\mathrm{Id}$ (and $\chi(x) = 1$), the equation for ψ^ε writes

$$i \partial_t u^\varepsilon + \frac{1}{2} \partial_y^2 u^\varepsilon = \mathcal{V}^\varepsilon u^\varepsilon + \varepsilon^{\alpha - 3/2} |u^\varepsilon|^2 u^\varepsilon,$$

where $\mathcal{V}^\varepsilon(t, y) = \frac{1}{\varepsilon}\left(\lambda(x(t) + \sqrt{\varepsilon}y) - \lambda(x(t)) - \sqrt{\varepsilon}\lambda'(x(t))y\right).$

Formally, $\mathcal{V}^\varepsilon(t, y) \xrightarrow[\varepsilon \to 0]{} \mathcal{V}(t, y) = \frac{1}{2}\lambda''(x(t))y^2$, which suggests the ansatz

$$\varphi^\varepsilon(t, x) = \frac{1}{\varepsilon^{1/4}} u\left(t, \frac{x - x(t)}{\sqrt{\varepsilon}}\right) e^{i(S(t) + \xi(t) \cdot (x - x(t))/\varepsilon}},$$

where the profile u satisfies different equation whether $\alpha > 3/2$ or not

$$i\partial_t u + \frac{1}{2}\partial_y^2 u = \begin{cases} 0 \text{ if } \alpha > 3/2, \\ \frac{1}{2}\lambda''(x(t))y^2 u + |u|^2 u \text{ if } \alpha = 3/2 \end{cases}; \quad u(0, y) = a(y). \quad (1)$$

In the following, we shall assume that we are in a nonlinear regime, i.e. $\alpha = 3/2$. Then, the existence and the properties of the solution to the nonlinear and non autonomous Schrödinger equation (1) is an issue by itself. It has been solved by Carles [3]: for $a \in \mathcal{S}$, there exists a unique global solution to (1). Besides, we have an exponential control of the growth of the momenta of u

In the matrix-valued setting, we also need to choose the eigenvectors that we will consider. We use the time dependent eigenvectors introduced by George Hagedorn in 1994 (see [8] and [4]): if d_j denotes the multiplicity of the eigenvalue λ_j, there exists a smooth orthonormal family $\left(\chi^\ell(t, x)\right)_{1 \le \ell \le d_1}$ such that for all t, $\left(\chi^\ell(t, x)\right)_{1 \le \ell \le d_1}$ spans the eigenspace associated to $\lambda_1 := \lambda$, $\chi^1(0, x) = \chi(x)$ and for $\ell \in \{1, \cdots, d_1\}$, the vector $\partial_t \chi^\ell(t, x) + \xi(t)\partial_x \chi^\ell(t, x)$ is in $\mathrm{Ker}\,(V(x) - \lambda(x)\mathrm{Id})^\perp$. Moreover, we have an exponential control in time of the growth of the derivatives of the vectors χ^ℓ.

3 Main Results

Theorem 1 (Adiabatic Theorem, [5]). *Let $a \in \mathcal{S}(\mathbb{R})$. Under (H1) and (H2), there exists $C > 0$ such that $w^\varepsilon(t, x) = \psi^\varepsilon(t, x) - \varphi^\varepsilon(t, x)\chi^1(t, x)$ satisfies*

$$\sup_{|t| \le C\log\log\frac{1}{\varepsilon}} \left(\|w^\varepsilon(t)\|_{L^2} + \|xw^\varepsilon(t)\|_{L^2} + \|\varepsilon\partial_x w^\varepsilon(t)\|_{L^2}\right) \xrightarrow[\varepsilon \to 0]{} 0.$$

We prove this theorem by energy estimates on the function $w^\varepsilon + \varepsilon g^\varepsilon$ where g^ε are correction terms belonging to the orthogonal to $\mathrm{Ker}\,(V(x) - \lambda(x)\mathrm{Id})$. These correction terms are here to compensate the component of the vector $r(t, x) = \partial_t \chi + \xi(t)\partial_x \chi$ on $\mathrm{Ker}\,(V(x) - \lambda(x)\mathrm{Id})^\perp$. We point out that if $V(x) = \lambda(x)\mathrm{Id}$, one can prove that the asymptotics holds up to (an analogue of) Ehrenfest time by using Strichartz estimates.

It is also possible to prove a nonlinear superposition result. We suppose

$$\psi_0^\varepsilon(x) = \varphi_1^\varepsilon(0, x)\chi_1(x) + \varphi_2^\varepsilon(0, x)\chi_2(x),$$

with $(x_1, \xi_1) \neq (x_2, \xi_2)$ when $\chi_1(x)$ and $\chi_2(x)$ are in the same eigenspace. We naturally associate with (x_j, ξ_j, χ_j), $j \in \{1, 2\}$ an ansatz φ_j^ε and we obtain the following result.

Theorem 2 (Nonlinear Superposition, [5]). *Set* $E_j = \frac{\xi_j^2}{2} + \tilde{\lambda}_j(x_j)$ *and suppose*

$$\Gamma = \inf_{x \in \mathbf{R}} \left| \tilde{\lambda}_1(x) - \tilde{\lambda}_2(x) - (E_1 - E_2) \right| > 0.$$

Then, there exists $C > 0$ *such that the function* $w^\varepsilon(t) = \psi^\varepsilon(t) - \varphi_1^\varepsilon \chi^1(t, x) - \varphi_2^\varepsilon \chi^2(t, x)$ *satisfies*

$$\sup_{t \leq C \log\log\frac{1}{\varepsilon}} \left(\|w^\varepsilon(t)\|_{L^2} + \|x w^\varepsilon(t)\|_{L^2} + \|\varepsilon \partial_x w^\varepsilon(t)\|_{L^2} \right) \underset{\varepsilon \to 0}{\longrightarrow} 0.$$

The proof of this theorem relies on energy estimates and a careful analysis of the nonlinear interactions between φ_1^ε and φ_2^ε. One can prove that interaction terms of the form $|\varphi_1^\varepsilon|^2 |\varphi_2^\varepsilon|$ are small provided the gap between the trajectory, $|x_1(t) - x_2(t)|$ is large enough. Then, the constant Γ allows to control the lengths of the time intervals where this gap is small.

4 Perspectives

Of course, the generalization of this result to higher dimension is a challenging problem. In that case, a proof by energy estimates is no longer possible and one needs to use Strichartz estimates for the matrix-valued Schrödinger operator $P(\varepsilon) = -\frac{\varepsilon^2}{2}\Delta + V(x)$. Such estimates are available in the literature under decay assumptions on $V(x)$ as $|x|$ goes to ∞. As far as we know, the existence of Strichartz estimates for the operator $P(\varepsilon)$ when $V(x)$ can have quadratic growth is an open question. It is also an issue to analyze the competition between the coupling by the non-linearity and the coupling produced by the potential in the case where $V(x)$ presents eigenvalue crossings. Indeed, in the linear setting, eigenvalue crossings are known to generate non adiabatic transitions between the modes. However, the analysis of crossings in the nonlinear regime has not yet been done.

References

1. D. Bambusi, S. Graffi, and T. Paul, Long time semiclassical approximation of quantum flows: a proof of the Ehrenfest time, Asymptot. Anal. 21 (1999), no. 2, p. 149–160.
2. W. Bao, Analysis and efficient computation for the dynamics of two-component Bose-Einstein condensates Stationary and Time-dependent Gross-Pitaevskii Equations (Contemporary Mathematics vol 473) p. 1–26 (2008).

3. R. Carles, Nonlinear Schrödinger equation with time dependent potential. (to appear in Commun. Math. Sci.).
4. R. Carles and C. Fermanian, Nonlinear coherent states and Ehrenfest time for Schrödinger equations, Commun. Math. Phys., 301 (2011), p. 443–472.
5. R. Carles and C. Fermanian, A Nonlinear Adiabatic Theorem for Coherent States, Nonlinearity, 24 (2011), no. 8, p. 2143–2164.
6. M. Combescure and D. Robert, Semiclassical spreading of quantum wave packets and applications near unstable fixed points of the classical flow, Asymptot. Anal. 14 (1997), no. 4, p. 377–404.
7. G. A. Hagedorn, Semiclassical quantum mechanics. I. The $h \to 0$ limit for coherent states, Comm. Math. Phys. 71 (1980), no. 1, p. 77–93.
8. G. A. Hagedorn, Molecular propagation through electron energy level crossings, Mem. Amer. Math. Soc. 111 (1994), no. 536, vi+130.
9. G. A. Hagedorn and A. Joye, Landau-Zener transitions through small electronic eigenvalue gaps in the Born-Oppenheimer approximation, Ann. Inst. H. Poincaré Phys. Théor. 68 (1998), no. 1, p. 85–134.
10. G. A. Hagedorn and A. Joye, Exponentially accurate semiclassical dynamics: propagation, localization, Ehrenfest times, scattering, and more general states, Ann. Henri Poincaré 1 (2000), no. 5, p. 837–883.
11. D.S. Hall, M.R. Matthews, J.R. Ensher, C.E. Wieman and E.A. Cornell, Dynamics of component separation in a binary mixture of Bose-Einstein condensates, Phys. Rev. Lett., 81, p. 1539–1542 (1998).
12. A. Martinez and V. Sordoni, Twisted pseudodifferential calculus and application to the quantum evolution of molecules. Mem. Am. Math. Soc. 200 (936) vi+82 (2009).
13. H. Spohn and S. Teufel, Adiabatic decoupling and time-dependent Born-Oppenheimer theory, Comm. Math. Phys. 224 (2001), no. 1, p. 113–132.
14. S. Teufel, Adiabatic perturbation theory in quantum dynamics, Lecture Notes in Mathematics, vol. 1821, Springer-Verlag, Berlin, 2003.

Adiabatic Limits and Related Lattice Point Problems

Yuri A. Kordyukov and Andrey A. Yakovlev

1 Preliminaries on Adiabatic Limits

Let $(M, \mathcal{F})$ be a closed foliated manifold endowed with a Riemannian metric g. Then we have a direct sum decomposition $TM = F \oplus H$ of the tangent bundle TM of M, where $F = T\mathcal{F}$ is the tangent bundle of $\mathcal{F}$ and $H = F^{\perp}$ is the orthogonal complement of F, and the corresponding decomposition of the metric: $g = g_F + g_H$. Consider the one-parameter family of Riemannian metrics on M,

$$g_\varepsilon = g_F + \varepsilon^{-2} g_H, \quad \varepsilon > 0,$$

and the corresponding Laplace-Beltrami operator Δ_ε. We are interested in the asymptotic behavior of the trace of the operator $f(\Delta_\varepsilon)$ for sufficiently nice functions f on $\mathbb{R}$, in particular, of the eigenvalue distribution function $N_\varepsilon(\lambda)$ of Δ_ε, as $\varepsilon \to 0$ (in the adiabatic limit).

In [4] (see also [2, 3, 5]), the first author proved an asymptotic formula for tr $f(\Delta_\varepsilon)$ in the case when the foliation $\mathcal{F}$ is Riemannian and the metric g is bundle-like. For particular examples of non-Riemannian foliations, such an asymptotic formula was proved by the second author in [11, 12] (see also a survey paper [6] for some historic remarks and references).

As the simplest example, one can consider the linear foliation $\mathcal{F}$ on the n-dimensional torus $\mathbb{T}^n = \mathbb{R}^n / \mathbb{Z}^n$ given by the leaves $L_x = x + F \mod \mathbb{Z}^n, x \in \mathbb{T}^n$, where F is a linear subspace of $\mathbb{R}^n$. Let g be the standard Euclidean metric on $\mathbb{T}^n$. The foliation $\mathcal{F}$ is Riemannian, and the metric g is bundle-like. In this case, the eigenvalue distribution function $N_\varepsilon(\lambda)$ of Δ_ε equals the number of integer points in

Y.A. Kordyukov (✉) · A.A. Yakovlev
Institute of Mathematics, Russian Academy of Sciences, 112 Chernyshevsky street, 450008, Ufa, Russia
e-mail: yurikor@matem.anrb.ru; yakovlevandrey@yandex.ru

D. Grieser et al. (eds.), *Microlocal Methods in Mathematical Physics and Global Analysis*, Trends in Mathematics, DOI 10.1007/978-3-0348-0466-0_6, © Springer Basel 2013

the ellipsoid $\{\xi \in \mathbb{R}^n : \sum_{j,\ell=1}^{n} g_\varepsilon^{j\ell} \xi_j \xi_\ell < \lambda/(2\pi)^2\}$. So we arrive at the following lattice point problem.

2   Lattice Point Problems

Let F be a p-dimensional linear subspace of $\mathbb{R}^n$ and $H = F^\perp$ the orthogonal complement of F with respect to the standard Euclidean inner product $(\cdot, \cdot)$ in $\mathbb{R}^n$, $p + q = n$. For any $\varepsilon > 0$, consider the linear transformation $T_\varepsilon : \mathbb{R}^n \to \mathbb{R}^n$ defined by

$$T_\varepsilon(x) = x, \text{ if } x \in F, \quad T_\varepsilon(x) = \varepsilon^{-1}x, \text{ if } x \in H.$$

Let S be a bounded open set with smooth boundary in $\mathbb{R}^n$. Put

$$n_\varepsilon(S) = \#(T_\varepsilon(S) \cap \mathbb{Z}^n), \quad \varepsilon > 0.$$

The problem is to study the asymptotic behavior of $n_\varepsilon(S)$ as $\varepsilon \to 0$. It appears that, in the general case, the leading term in the asymptotic formula for $n_\varepsilon(S)$ as $\varepsilon \to 0$ was unknown. In a slightly different context, this problem was studied in considerable detail in [9, 10] (see also the references therein).

Let $\Gamma = \mathbb{Z}^n \cap F$. Γ is a free abelian group. Denote by $r = \operatorname{rank} \Gamma \leq p$ the rank of Γ. Let V be the r-dimensional subspace of $\mathbb{R}^n$ spanned by the elements of Γ. Observe that Γ is a lattice in V. Let $\Gamma^* \subset V$ denote the dual lattice to Γ: $\Gamma^* = \{\gamma^* \in V : (\gamma^*, \Gamma) \subset \mathbb{Z}\}$. For any $x \in V$, denote by P_x the $(n - r)$-dimensional affine subspace of $\mathbb{R}^n$, passing through x orthogonal to V.

Theorem 1 ([7]). *Under the current assumptions, we have*

$$n_\varepsilon(S) = \frac{\varepsilon^{-q}}{\operatorname{vol}(V/\Gamma)} \sum_{\gamma^* \in \Gamma^*} \operatorname{vol}_{n-r}(P_{\gamma^*} \cap S) + R_\varepsilon(S),$$

where the remainder $R_\varepsilon(S)$ satisfies the estimate

$$R_\varepsilon(S) = O(\varepsilon^{\frac{1}{p-r+1} - q}), \quad \varepsilon \to 0.$$

Theorem 2 ([7, 8]).

(1) If, for any $x \in F$, the intersection $\{x + H\} \cap S$ is strictly convex, then

$$R_\varepsilon(S) = O(\varepsilon^{\frac{2q}{q+1+2(p-r)} - q}), \quad \varepsilon \to 0.$$

(2) If, for any $x \in F$, the intersection $\{x + V^\perp\} \cap S$ is strictly convex, then

$$R_\varepsilon(S) = O(\varepsilon^{\frac{2q}{n-r+1} - q}), \quad \varepsilon \to 0.$$

In [8], we also study the average remainder estimates, where the average is taken over rotations of the domain S by orthogonal transformations in $\mathbb{R}^n$.

3 Applications to Adiabatic Limits

As a straightforward consequence of Theorem 2, we obtain a more precise estimate for the remainder in the asymptotic formula of [4] in the above mentioned case when $\mathcal{F}$ is a linear foliation on $\mathbb{T}^n$ and g is the standard Euclidean metric.

Theorem 3 ([7]). *For $\lambda > 0$, we have, as $\varepsilon \to 0$,*

$$
N_\varepsilon(\lambda) = \varepsilon^{-q} \frac{\omega_{n-r}}{\text{vol}(V/\Gamma)} \sum_{\gamma^* \in \Gamma^*} \left(\frac{\lambda}{4\pi^2} - |\gamma^*|^2 \right)^{(n-r)/2} + O(\varepsilon^{\frac{2q}{n-r+1} - q}),
$$

where ω_{n-r} is the volume of the unit ball in $\mathbb{R}^{n-r}$.

4 Some Open Problems

1. *Prove an asymptotic formula for* tr $f(\Delta_\varepsilon)$ *when $\mathcal{F}$ is an arbitrary foliation.*
   The case when $\mathcal{F}$ is given by the fibers of a fibration over a compact manifold and the metric g is not bundle-like, is already quite interesting.
2. *Prove a complete asymptotic expansion for the heat trace* tr $e^{-t\Delta_\varepsilon}$ *as $\varepsilon \to 0$ (even if the metric g is bundle-like).*
3. *Study the adiabatic limits of more complicated spectral invariants like the eta-invariant, the analytic torsion etc. (even if the metric g is bundle-like).*
   Here the extension of the Mazzeo-Melrose result on small eigenvalues in the adiabatic limit and spectral sequences to Riemannian foliations [1] might be useful.
4. *Study the remainder estimates for $N_\varepsilon(\lambda)$.*
5. *Continue the study of the remainder $R_\varepsilon(S)$, depending on geometry of a domain S and properties of F and H.*

Acknowledgements Supported by the Russian Foundation of Basic Research (09-01-00389 and 12-01-00519).

References

1. J. Álvarez López, Yu. A. Kordyukov, *Adiabatic limits and spectral sequences for Riemannian foliations*, Geom. and Funct. Anal. **10** (2000), 977–1027.

2. Yu. A. Kordyukov, *Quasiclassical asymptotics of the spectrum of elliptic operators in a foliated manifold.* Math. Notes **53** (1993), no. 1–2, 104–105.
3. Yu. A. Kordyukov, *On the quasiclassical asymptotics of the spectrum of hypoelliptic operators on a manifold with foliation.* Funct. Anal. Appl. **29** (1995), no. 3, 211–213.
4. Yu. A. Kordyukov, *Adiabatic limits and spectral geometry of foliations*, Math. Ann. **313** (1999), 763–783.
5. Yu. A. Kordyukov, *Semiclassical spectral asymptotics on foliated manifolds*, Math. Nachr. **245** (2002), 104–128.
6. Yu. A. Kordyukov, A. A. Yakovlev, *Adiabatic limits and the spectrum of the Laplacian on foliated manifolds*, C^*-algebras and elliptic theory. II, Trends in Mathematics, Birkhäuser, Basel, 2008, 123–144.
7. Yu. A. Kordyukov, A. A. Yakovlev, *Integer points in domains and adiabatic limits*, Algebra i Analiz, **23** (2011), no. 6, 80–95; preprint arXiv:1006.4977.
8. Yu. A. Kordyukov, A. A. Yakovlev, *The number of integer points in a family of anisotropically expanding domains*, in preparation.
9. Nikichine, N. A.; Skriganov, M. M., *Nombre de points d'un réseau dans un produit cartésien de domaines convexes*, C. R. Acad. Sci. Paris Sér. I Math. **321** (1995), no. 6, 671–675.
10. Nikishin, N. A.; Skriganov, M. M.. *On the distribution of algebraic numbers in parallelotopes*, St. Petersburg Math. J. **10** (1999), no. 1, 53–68
11. A. A. Yakovlev, *Adiabatic limits on Riemannian Heisenberg manifolds*, Sb. Math. **199** (2008), no. 1–2, 307–318.
12. A. A. Yakovlev, *Asymptotics of the spectrum of the Laplace operator on Riemannian Sol-manifolds in the adiabatic limit*, Sib. Math. J. **51** (2010), no. 2, 370–382.

The Effective Hamiltonian in Curved Quantum Waveguides and When It Does Not Work

David Krejčiřík and Helena Šedivákova

We are concerned with the singular operator limit for the Dirichlet Laplacian in a three-dimensional curved tube (cf. Fig. 1) when its cross-section shrinks to zero. The tube Ω_ε is constructed by translating and rotating the cross-section along a spatial (bounded, unbounded or semibounded) curve Γ and the limit is realized by scaling a fixed cross-section ω by a small positive number ε. Geometrically, Ω_ε collapses to Γ as $\varepsilon \to 0$. We are interested in how and when the three-dimensional Laplacian $-\Delta_D^{\Omega_\varepsilon}$ can be approximated by a one-dimensional operator H_{eff} on the curve.

We start with some more or less obvious observations.

1. Since we deal with unbounded operators, the convergence of $-\Delta_D^{\Omega_\varepsilon}$ to H_{eff} is understood through a convergence of their resolvents.
2. The Dirichlet boundary conditions imply that the spectrum of $-\Delta_D^{\Omega_\varepsilon}$ explodes as $\varepsilon \to 0$. It is just because the first eigenvalue of the (two-dimensional) Dirichlet Laplacian in the scaled cross-section $\varepsilon\omega$ equals $\varepsilon^{-2}E_1$, where E_1 is the first eigenvalue of the Dirichlet Laplacian in the fixed cross-section ω. Hence, a normalization $-\Delta_D^{\Omega_\varepsilon} - \varepsilon^{-2}E_1$ is in order to get a non-trivial limit.
3. Finally, since the configuration spaces Ω_ε and Γ have different dimensions, a suitable identification of respective Hilbert spaces of $-\Delta_D^{\Omega_\varepsilon}$ and H_{eff} is required. This is achieved by considering H_{eff} as acting on the subspace of $L^2(\Omega_\varepsilon)$ spanned by functions of the form $\phi \otimes \mathcal{J}_1$ on $\Gamma \times \omega$, where $\mathcal{J}_1$ denotes the positive normalized eigenfunction of $-\Delta_D^\omega$ corresponding to E_1.

D. Krejčiřík (✉)
Nuclear Physics Institute ASCR, 25068, Řež, Czech Republic
e-mail: krejcirik@ujf.cas.cz

H. Šedivákova
Faculty of Nuclear Sciences and Physical Engineering, Czech Technical University, Břehová 7, 11519, Praha 1, Czech Republic
e-mail: sedivakova@ujf.cas.cz

D. Grieser et al. (eds.), *Microlocal Methods in Mathematical Physics and Global Analysis*, Trends in Mathematics, DOI 10.1007/978-3-0348-0466-0_7, © Springer Basel 2013

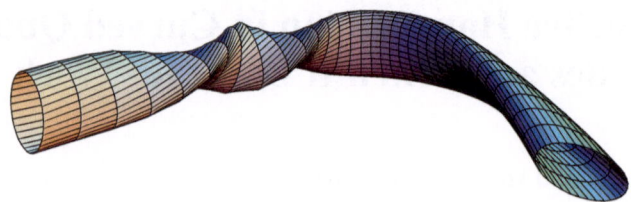

Fig. 1 An example of a waveguide of elliptical cross-section. Twisting and bending are demonstrated on the *left* and *right* part of the picture, respectively

It is well known how the limiting operator looks like [2, 4, 6]. We have

$$- \Delta_D^{\Omega_\varepsilon} - \varepsilon^{-2} E_1 \xrightarrow[\varepsilon \to 0]{} H_{\text{eff}} := -\Delta_D^\Gamma - \frac{\kappa^2}{4} + C_\omega \left(\dot{\theta}_F - \tau \right)^2 \qquad (1)$$

in a resolvent sense, where κ and τ denote respectively the curvature and torsion of Γ, θ_F is an angle function defining the rotation of $\varepsilon\omega$ with respect to the Frenet frame of Γ and $C_\omega := \|\partial_\varphi \mathcal{J}_1\|_{L^2(\omega)}$, with ∂_φ denoting the angular derivative. The potential of H_{eff} clearly consists of two competing terms, which represent the opposite effects of bending and twisting in quantum waveguides, cf. [5].

The question we would like to address here is about the optimal regularity conditions under which the effective approximation (1) holds. We are motivated by the fact that the known existing results mentioned above do not cover physically interesting curves with discontinuous curvature and without Frenet frame. Indeed, the latter is a standard hypothesis in the literature in order to construct the waveguide. However, the Frenet frame exists only for curves of class C^3 with nowhere vanishing curvature κ.

Furthermore, the Γ-convergence method of [2, 4], which seems to work under less restrictive regularity once the technical difficulty of the non-existence of the Frenet frame is overcome, implies only (unless the waveguide is bounded [4]) a strong-resolvent convergence for (1) and does not provide any information about the convergence rate. Our goals are thus as follows:

1. Impose no unnatural hypothesis about the reference curve Γ, include curves which are merely of class $W_{\text{loc}}^{2,\infty}$ and which do not possess Frenet frame.
2. Use operator methods instead of the Γ-convergence, establish (1) in the norm-resolvent sense and get a control on the convergence rate.

Ou strategy how to achieve these goals is based on the following ideas:

1. Use the frame defined by the parallel transport instead of the Frenet frame. This alternative frame is known to exist for any curve of class C^2, cf. [1]. We generalize the construction to the curves of class $W_{\text{loc}}^{2,\infty}$.
2. Work exclusively with the quadratic forms associated with the operators.

Even if one implements the above ideas, the standard operator approach to the thin-width limit in quantum waveguides (see, e.g., [3]) still requires certain differentiability of curvature κ (which is just bounded under our hypotheses). To see it, we briefly recall the standard strategy.

First, one uses curvilinear coordinates, which induce the unitary transform $U_1 : L^2(\Omega_\varepsilon) \to L^2(\Gamma \times \omega, \varepsilon^2 h_\varepsilon(s, t) \, ds \, dt)$, with the Jacobian

$$h_\varepsilon(\cdot, t) := 1 - \varepsilon \, t_1 \, [k_1 \cos \theta - k_2 \sin \theta] - \varepsilon \, t_2 \, [k_1 \sin \theta + k_2(s) \cos \theta]. \quad (2)$$

Here k_1, k_2 are curvature functions computed with respect to our relatively parallel frame and θ is an angle function defining the rotation of the cross-section $\varepsilon\omega$ with respect to the frame. We have $\kappa^2 = k_1^2 + k_2^2$ and, if the Frenet frame exists, our frame is rotated with respect to the Frenet frame by the angle given by a primitive of torsion τ. Consequently, in our more general setting, the difference $\dot{\theta}_F - \tau$ in (1) is to be replaced by $\dot{\theta}$.

Second, to recover the curvature term in the effective potential of (1), one also performs the unitary transform $U_2 : L^2(\Gamma \times \omega, \varepsilon^2 \, h_\varepsilon(s, t) \, ds \, dt) \to L^2(\Gamma \times \omega)$ generated by $\psi \mapsto \varepsilon \sqrt{h_\varepsilon}\psi$. However, this transform does not leave the form domain $W^{1,2}(\Gamma \times \omega)$ invariant if k_1, k_2 are not smooth in a suitable sense.

This last difficulty is overcome by the following trick:

3. Replace the curvature functions in (2) by their ε-dependent mollifications ($\mu \in \{1, 2\}$)

$$k_\mu^\varepsilon(s) := \frac{1}{\delta(\varepsilon)} \int_{s - \frac{\delta(\varepsilon)}{2}}^{s + \frac{\delta(\varepsilon)}{2}} k_\mu(\xi) \, d\xi,$$

where δ is a continuous function such that $\varepsilon^{-1}\delta(\varepsilon)$ diverges as $\varepsilon \to 0$.

Then everything works very well because the longitudinal derivative of the mollified h_ε involves the terms $\varepsilon \dot{k}_\mu^\varepsilon$ which vanish in the limit $\varepsilon \to 0$, even if $\dot{k}_\mu^\varepsilon$ diverge.

Our main result can be informally stated as follows:

Theorem 1. *Let $\Gamma \in W_{loc}^{2,\infty}$ with $\kappa \in L^\infty$ and $\theta \in W_{loc}^{1,\infty}$ with $\dot{\theta} \in L^\infty$. Then* (1) *holds in the norm-resolvent sense provided*

$$\sigma(\varepsilon) := \sum_{f = k_1, k_2, \dot{\theta}} \sup_{n \in \mathbb{Z}} \sqrt{\sup_{|\xi| < \frac{\delta_f(\varepsilon)}{2}} \int_n^{n+1} |f(s + \xi) - f(s)|^2 \, ds} \quad (3)$$

tends to zero as $\varepsilon \to 0$.

Our sufficient condition for the validity of the effective approximation looks horrible. But it is actually not so bad. It is easy to verify that it covers all the known results, and much more. In particular, it holds if all the representants of f are either Lipschitz, or just uniformly continuous, or square integrable (i.e., always whenever Γ is bounded), or periodic, etc. Furthermore, the quantity $\sigma(\varepsilon)$ together with $\varepsilon \|\dot{k}_\mu^\varepsilon\|_\infty$ determines the decay rate of the convergence (1).

An open problem is to get rid of the hypothesis $\sigma(\varepsilon) \to 0$ as $\varepsilon \to 0$, if possible.

References

1. R. L. Bishop, *There is more than one way to frame a curve*, The American Mathematical Monthly **82** (1975), 246–251.
2. G. Bouchitté, M. L. Mascarenhas, and L. Trabucho, *On the curvature and torsion effects in one dimensional waveguides*, ESAIM: Control, Optimisation and Calculus of Variations **13** (2007), 793–808.
3. P. Duclos and P. Exner, *Curvature-induced bound states in quantum waveguides in two and three dimensions*, Rev. Math. Phys. **7** (1995), 73–102.
4. C. R. de Oliveira, *Quantum singular operator limits of thin Dirichlet tubes via Γ-convergence*, Rep. Math. Phys. **66** (2010) 375–406.
5. D. Krejčiřík, *Twisting versus bending in quantum waveguides*, Analysis on Graphs and its Applications, Cambridge, 2007 (P. Exner et al., ed.), Proc. Sympos. Pure Math., vol. 77, Amer. Math. Soc., Providence, RI, 2008, pp. 617–636; arXiv:0712.3371 [math-ph].
6. J. Lampart, S. Teufel, and J. Wachsmuth, *Effective Hamiltonians for thin Dirichlet tubes with varying cross-section*, Mathematical Results in Quantum Physics, September, 2010, xi+274 p.; World Scientific, Singapore, 2011, pp. 183–189; arXiv:1011.3645 [math-ph].

The Adiabatic Limit of the Laplacian on Thin Fibre Bundles

Jonas Lampart and Stefan Teufel

We consider the time dependent Schrödinger equation on a fibre bundle in the adiabatic limit. We allow the fibres to have a boundary, in which case we impose Dirichlet conditions. In particular this allows us to understand, in a general setting, results obtained in the context of quantum waveguides (see the review [2]), in which case the bundle is usually a solid cylinder or a square. In order to extract such results from our effective operator one has to embed the total space into $\mathbb{R}^n$ and expand the induced metric. The leading order term will be a Riemannian submersion. When calculating the effective operator the additional corrections due to the metric need to be taken into account. The resulting operator can then be analysed to obtain e.g. expansions for the eigenvalues.

One of the virtues of the approach is a scaling that gives results for energies that are infinite in many of the settings treated in the literature. More precisely, Theorem 1 gives an approximation for finite energies when the energies of the fibre dynamics are independent of the scaling parameter. In comparison the often considered scaling in which these energies go to infinty as the fibres get thin only yields results for small energies above the ground state (after substracting the increasing energy of the fibre ground state).

We use methods of adiabatic perturbation theory (see [3] for a comprehensive presentation). The basic idea is that the separation of scales between the base and the fibre leads to a decoupling of the corresponding dynamics. This means that for special initial conditions (one may think of eigenstates of the fibre dynamics) the dynamics in the fibre direction will be trivial for long times. One can thus decompose the total problem into a set of simpler problems, one for every such initial condition. Each of these has dynamics only in directions on the base and is governed by an effective equation.

J. Lampart (✉) · S. Teufel
Universität Tübingen, Mathematisches Institut, Auf der Morgenstelle 10, 72076, Tübingen, Germany
e-mail: jola@maphy.uni-tuebingen.de; stefan.teufel@uni-tuebingen.de

D. Grieser et al. (eds.), *Microlocal Methods in Mathematical Physics and Global Analysis*, Trends in Mathematics, DOI 10.1007/978-3-0348-0466-0_8, © Springer Basel 2013

These methods were applied in a differential geometric setting to the problem of constraints in quantum mechanics [4].

Let $F \to M \xrightarrow{\pi} B$ be a smooth fibre bundle with a Riemannian submersion metric $g = g_{F_x} \oplus \pi^*h$ (see [1, Chap. 2] for a discussion of basic properties). By the adiabatic limit we mean the asymptotic limit as $\varepsilon \to 0$ for the scaled family of metrics $g_\varepsilon = g_{F_x} \oplus \varepsilon^{-2}\pi^*h$. We study this limit for the time dependent Schrödinger equation

$$i\partial_t \psi = -\Delta_{g_\varepsilon} \psi. \tag{1}$$

Δ_{g_ε} is self adjoint on the Dirichlet domain $H^2(M) \cap H_0^1(M)(=H^2(M)$ if $\partial M = \emptyset)$.

It can easily be seen (for example using the quadratic form) that the Laplacian decomposes into

$$\Delta_{g_\varepsilon} = \mathrm{tr}_{TF}\nabla^2 + \varepsilon^2\left(\mathrm{tr}_{NF}\nabla^2 - \eta\right) =: \Delta_{F_x} + \varepsilon^2\Delta_h \tag{2}$$

where η is the mean curvature vector of the fibres (in the metric $g_{\varepsilon=1}$). We interpret these terms in the following way:

- For every x the Laplacian of the fibre Δ_{F_x} is a bounded linear operator from its domain to $L^2(F_x)$. It can thus be viewed as a section of a bundle over B: Let $L^2(F;\pi)$ and D be vector bundles induced by π (with fibres $L^2(F)$ and $H^2(F_x) \cap H_0^1(F_x)$ respectively). The fibre Laplacian is precisely a section of the bundle of continuous linear maps $\mathcal{L}\left(D, L^2(F;\pi)\right)$ between both.
- Derivation in the horizontal direction formally defines a connection ∇^h on $L^2(F;\pi)$. $-\Delta_h$ can be identified as the Laplacian $\nabla^{h*}\nabla^h$ of this connection.

From this point of view the operator $\Delta_{g_\varepsilon} = \Delta_{F_x} + \varepsilon^2\Delta_h$ fits nicely into the framework of adiabatic perturbation theory, where one might call it "fibred over B". Now we proceed by noting that for every $x \in B$ the spectrum of Δ_{F_x} is discrete, of finite multiplicity and independent of ε. As it depends continuously on x it has a band structure i.e. one can choose continuous functions $E(x)$ that are eigenvalues for every x.

Let E be such a band and $P(x)$ its spectral projection, i.e.

$$\forall x: \quad -\Delta_{F_x}P(x) = E(x)P(x). \tag{3}$$

Since $P(x)$ is a bounded linear map on $L^2(F_x)$, P is a section of $\mathcal{L}(L^2(F;\pi))$. If E is separated from the rest of the spectrum then the dimension of ran P is constant and $PL^2(F;\pi)$ is a finite-rank subbundle that is locally spanned by eigenfunctions of Δ_{F_x}.

One can show that $PL^2(M)$ is left invariant by the dynamics up to errors of order ε by estimating $[\Delta_{g_\varepsilon}, P] = [\varepsilon^2\Delta_h, P]$ using the following:

Lemma 1. *Let M, B and π be of bounded geometry.[1] Let E be bounded and uniformly separated.*
Then $\left\| [\varepsilon \nabla_X^h, P] P \right\|_{\mathcal{L}(L^2(M))} \leq C \varepsilon$ for every bounded X.

This allows us to describe the leading order dynamics (with initial conditions in $PL^2(M)$) by the effective operator $P \Delta_{g_\varepsilon} P$ on the L^2-sections of our finite-rank bundle $PL^2(F; \pi)$. However we aim for more precision because:

1. Horizontal distances increase as $\varepsilon \to 0$, so with kinetic energies of order one it takes times of order $1/\varepsilon$ for global effects to occur, and
2. The spacing of the eigenvalues of Δ_{g_ε} decreases and an error of order ε might not be enough to distinguish them.

Therefore we need better approximations to understand the dynamics and completely resolve the spectrum. For this purpose we construct a corrected projection $P^\varepsilon = P + \mathcal{O}(\varepsilon)$ (that is no longer a fibrewise operator), for which $[\Delta_{g_\varepsilon}, P] P = \mathcal{O}(\varepsilon^N)$ holds, and an intertwining unitary $U^\varepsilon : P^\varepsilon L^2(M) \to PL^2(M)$ that allows us to define the effective operator on sections of $PL^2(F; \pi)$ (see [4] for a detailed exposition of the technique).

Theorem 1. *Under the assumptions of the lemma there exist*

- *A projection $P^\varepsilon \in \mathcal{L}\left(L^2(M)\right)$,*
- *A unitary operator $U^\varepsilon \in \mathcal{L}\left(L^2(M)\right)$ that maps $P^\varepsilon L^2(M) \to PL^2(M)$,*
- *A self-adjoint operator $(H_{\mathrm{eff}}, D_{\mathrm{eff}})$ on $L^2(PL^2(F; \pi))$,*

such that

$$\left\| \left(e^{-iH^\varepsilon t} - U^{\varepsilon *} e^{-iH_{\mathrm{eff}} t} U^\varepsilon \right) P^\varepsilon \chi_{(-\infty, E_{\max})}(\Delta_{g_\varepsilon}) \right\|_{\mathcal{L}(L^2(M))} \leq C \varepsilon^N t \qquad (4)$$

for all $t \geq 0$ and $E_{\max} < \infty$;
If λ is an eigenvalue of H_{eff} then there is a ball B of radius $C \varepsilon^N$ around λ such that $B \cap \sigma\left(\Delta_{g_\varepsilon}\right) \neq \emptyset$.

The effective operator is given by

$$H_{\mathrm{eff}} = P U^\varepsilon \Delta_{g_\varepsilon} U^{\varepsilon *} P = -\varepsilon^2 \Delta_B + E(x) + \mathcal{O}(\varepsilon^2), \qquad (5)$$

where $-\Delta_B$ is the Laplacian of the induced connection $\nabla^B = P \nabla^h P$. The higher order corrections can be computed explicitly from the construction of P^ε.

---

[1] When there is no boundary we require that M, B be of bounded geometry and π have bounded derivatives. In the case with boundary there is a similar condition. Both are trivially satisfied if M is compact.

References

1. Peter B. Gilkey, John V. Leahy, and Jeonghyeong Park, *Spectral geometry, Riemannian submersions, and the Gromov-Lawson conjecture*, Studies in Advanced Mathematics, Chapman & Hall/CRC, 1999.
2. D. Grieser, *Thin tubes in mathematical physics, global analysis and spectral geometry*, Analysis on Graphs and its Applications, Proceedings of Symposia in Pure Mathematics, vol. 77, 2008.
3. Stefan Teufel, *Adiabatic perturbation theory in quantum dynamics*, Lecture Notes in Mathematics, Springer, 2003.
4. Jakob Wachsmuth and Stefan Teufel, *Effective Hamiltonians for constrained quantum systems*, arXiv:0907.0351 [math-ph], 2009.

Adiabatic Limit with Isolated Degenerate Fibres

Richard B. Melrose

In this talk I want to remind you of the setup for adiabatic limit and then to discuss various generalizations of it. I will try to show how problems with singular or degenerate fibres can be treated and how these are related to gluing problems and to the behaviour at the boundary of certain geometric moduli spaces.

1 Adiabatic Limit

Recall that the basic set up of an adiabatic limit corresponds to a fibration of manifolds – a submersion between compact connected manifolds $\phi : X \longrightarrow Y$ with typical fibre Z. On the total space X one can consider a family of 'adiabatic metrics' $g_t = \phi^* h + t^2 \mu$ where h is a metric on the base (it could also depend on t) and μ is a smooth symmetric 2-cotensor on the fibres which restricts to be positive definite, and hence a metric, on the fibres. In fact I think it is more natural to consider a family of metrics such as $t^{-2} g_t$ where the fibres are of more-or-less constant size and fixed vectors in the base get 'large' which means the base is 'slow' (hence the term adiabatic). I believe Witten [4] was the first to consider global analysis related to such metrics when he examined the behaviour of the eta invariant for the particular case of a fibration over a circle.

Note that this setting is more general than a Riemannian submersion, which corresponds to the case that μ has rank exactly equal to the dimension of the fibres at each point and I will mention some other possible generalizations below.

For $t > 0$ nothing much is happening, one just has a smooth family of metrics and t is simply a parameter. On the other hand one can view the singular limit at $t = 0$ as imposing a 'geometry'. Then t is no longer a true parameter but should

R.B. Melrose (✉)
Department of Mathematics, MIT, Cambridge, MA, USA
e-mail: rbm@math.mit.edu

D. Grieser et al. (eds.), *Microlocal Methods in Mathematical Physics and Global Analysis*,   37
Trends in Mathematics, DOI 10.1007/978-3-0348-0466-0_9, © Springer Basel 2013

be included in the analysis, so instead of X consider the space $X \times [0, 1]$ where the iterval corresponds to t. The basic object to consider is the space of smooth vector fields on $X \times [0, 1]$

$$\mathcal{V}_a(X) = \{V; Vt = 0, \ V \text{ tangent to } \phi^{-1}(y) \text{ at } t = 0\}. \tag{1}$$

These can also be identified as the smooth t-dependent vector fields on X such that $g_t(V) = O(t^2)$. They form a Lie algebra and module over $\mathcal{C}^\infty(X \times [0, 1])$ which implies that it makes sense to consider (for any vector bundles W_i over X) the filtered 'enveloping algebra' $t^l \operatorname{Diff}_a^k(X; W_1, W_2)$ given by the sums of products of up to k elements of this Lie algbra with an overall factor of t^l. The Laplacian of g_t, $\Delta_t \in t^{-2} \operatorname{Diff}_a^2(X; \Lambda^*)$.

Some year ago, Rafe Mazzeo and I considered the invertibility properties of the Laplacian on forms on X in this context. Let me use these results to illustrate the general behaviour of adiabatic problems. Roughly speaking the Laplacian is similar to the product case when we might have $\Delta_t = \Delta_Y + t^{-2}\Delta_Y$ but where the twisting does have an effect. In terms of the existence of a generalized inverse – which exists for $t > 0$, so we are really considering the regularity theory as $t \downarrow 0$ – there always exists a two-sided parametrix P_t such that

$$\Delta_t P_t - \operatorname{Id}, \ P_t \Delta_t - \operatorname{Id} \text{ are uniformly bounded on } L^2 \text{ and of finite rank}$$
$$\text{with } \|P_t\|_{L^2} \text{ bounded as } t \downarrow 0. \tag{2}$$

Of course one can say a lot more than this. In fact the generalized Hodge inverse Q_t is such that

$$\Delta_t Q_t - \operatorname{Id} = Q_t \Delta_t - \operatorname{Id} = \Pi(t), \ \Pi(t)^2 = \Pi(t) = \Pi^*(t) \text{ for } t > 0$$
$$\text{with } \Pi(t)u = u \iff \Delta_t u = 0, \ t > 0, \ \|Q_t\|_{L^2} \le C t^{-2k}. \tag{3}$$

2 Uniform Degeneracy

Although the 'main' solvability issue above appears to be the invertibility of the fibre Laplacian. Really, it is not quite this which is involved, or rather it is a little more than that. Namely what we actually get is a *suspended* Laplacian. This is the 'adiabatic symbol'. Note that the 'semiclassical' case is when ϕ is the identity. Then the adiabatic, or semiclassical, symbol is the 'full symbol' of the Laplacian. In case of a general fibration ϕ it is a bundle, over T^*Y, of operators. For each point in the base we get a differential operator (on some form bundle) over the space $T_y Y \times Z_y$, where Z_y is the fibre above y. This is a second order elliptic differential operator and is translation-invariant in $T_y Y$. We can take the Fourier transform and hence get a differential operator on Z_y which is polynomial in $T_y^* Y$. The main issue is then is the invertibility of this 'suspended' family (suspended in the topological sense of having some Euclidean variables added). For an adiabatic metric such as described above this turns out to be straightforward, the null space can be decomposed (over forms from the base) in terms of the fibre null spaces which form an ordinary vector bundle, the forms on Y twisted by the flat bundles of fibre harmonic forms.

So for the Laplacian on forms this model operator can never be fully invertible – the constants in degree 0 always intervene. For other similar problems it can. For instance if the fibres are manifolds with boundary and one considers the Dirichlet boundary condition, then the adiabatic model operators are fully invertible and in consequence the full Laplacian is also invertible uniformly:-

$$\|\Delta_{t,\text{Dir}}^{-1}\|_{L^2} \leq C t^2 \text{ as } t \downarrow 0. \tag{4}$$

Then a more interesting question, touched on below, is the behaviour of the lowest eigenvalue and eigenstate.

So the case of no boundary is 'uniformly degenerate' because these fibre operators are non-invertible in a uniform way and the invertibility properties described above depend on the invertibility (up to finite rank) of the induced Laplacian on these bundles. There is a second 'level' of solvability corresponding to the formal term 'Δ_Y' and indeed there is an induced Laplacian on the bundle, over Y, of null spaces at the first level. Correspondingly, the spectrum of Δ_t has three main 'pieces' as $t \downarrow 0$. There is a 'big' part which corresponds to non-zero eigenvalues of the fibre Laplacian, these behave like t^{-2} – for the Dirichlet case this is everything. There is a 'finite but non-zero' part which tends to non-zero constants – these correspond to the non-zero eigenvalues of Δ_Y. The third part arises from eigenvalues which tend to zero with t; it is finite-dimensional. This can be further analysed corresponding to the order of vanishing and leads to the various estimates above; the best k in (3) corresponds to the level at which the Leray-Serre spectral sequence for the cohomology of X degenerates – anything smaller than this is in the null space for all t.

Next let me note that the adiabatic limit in the form discussed above does have direct geometric applications. One such is in the work of Joel Fine [2] on the existence of Kähler metrics of constant mean curvature. I do not have time to get into the details of this, but the set up is as above, where ϕ is a 'fibration' in the case of X a compact complex surface (so having real dimension 4) and Y is a Riemann surface, so is Z. The first thing to understand is that in the complex, holomorphic, world (holomorphic) submersions are not (holomorphic) fibrations in the obvious sense, since in general the complex structure on the fibres varies – they are diffeomorphic in the real sense, but are not biholomorphical. Here this happens, except in low genus. In his thesis (some years ago) Fine shows how to 'make' a Kähler metric with constant scalar curvature on the total space from the 'obvious' metrics on base and fibre. Namely, since these are Riemann surfaces, there are metrics of constant curvature on them – including a smooth family of metrics on the fibres. Making (non-trivial) computations in the world of Kähler potentials, Fine shows how to construct a constant curvature adiabatic Kähler metric up to infinite order, i.e. in the sense of Taylor series (so the adiabatic family is Kähler and the scalar curvature is asymptotically just a function of t). Then the perturbation to an exact solution is an application of the implicit function using the invertibility properties of an operator (fourth order) like the Laplacian above. The message to take from this is that if one can solve an adiabatic problem to infinite order one can likely solve it exactly (although this is not the case for superficially similar similar problems, such as Witten's Morse complex).

3 Degenerate Fibres: Analytic Case

One natural question is what happens to the adiabatic limit if the form of the metric is generalized. This is related to issues below and there are several levels of 'relaxation' of the conditions. One might first consider perturbations of first order in t which include cross terms between base and fibre such as $t\,dy \otimes dz$. Note that it does not really make sense to want these to have coefficients which only depend on the base variables. The effect of adding such a term can be quite dramatic, since it can change the invertibility properties of the adiabatic model operators although this is still an elliptic suspended family – the null spaces may no longer form a bundle, as they may not be smooth over the base. If the failure to be smooth is itself reasonably smooth, as discussed below, then something can be done. I don't really know what happens to the null space in the general case – if anyone wants to try to work it out they are welcome to try and I am interested to discuss it!

Another possible 'perturbation' is to replace h by a basic tensor – that is, to let its coefficients vary on the fibre. This is actually less of a problem than the introduction of cross terms and leads to a very similar structure, but the details have not been written up as far as I know.

Next consider the effect of adding a potential, for simplicity in the case of the Laplacian on functions. The simplest case is when the potential is real and non-negative. If it is non-zero on any fibre then the model on that fibre is invertible. The opposite extreme to uniform degeneracy is when V vanishes on isolated fibres and has non-zero Hessian (in the base variables) at every point on those fibres

$$0 \leq V \in \mathcal{C}^{\infty}(X),$$

$$\exists \{y_1, \ldots, y_k\} \subset Y \text{ s.t. } V(x) = 0 \Longleftrightarrow x \in \bigcup_j Z_{y_j}, \tag{5}$$

$$\operatorname{Hess}_y V(y_j) > 0.$$

Then $\Delta_t + V$ has an inverse with a weaker (optimal) bound than in a case such as (4) when the model operators are invertible

$$\|(\Delta_t + V)^{-1}\|_{L^2} \leq Ct \tag{6}$$

In this case there are additional model operators at the singular fibres which are harmonic oscillators.

4 Eigenvalues for Triangles

One geometric setting closely related to the perturbation by a potential with isolated minima is the vertical collapse of a manifold with boundary. For example, if one takes a region in the plane between two 2π-periodic smooth curves, $\Omega = \{(x, y) \in \mathbb{R}^2; L(x) < y < U(x)\}$, and 'collapses' it by vertical scaling to $\Omega_t = \{(x, y) \in \mathbb{R}^2; L(x) < y/t < U(x)\}, 0 < t \leq 1$, then consider the behaviour of the eigenvalues for the Dirichlet, or Neumann, problem. For the Dirichlet problem the smallest eigenvalue has an asymptotic expansion related to that above, in particular there are harmonic oscillator models, provided $U(x) - L(x)$ only has non-degenerated maxima.

The general problem of the behaviour of the eigenvalues for the Dirichlet problem for triangles as functions on moduli space remains open and certainly there is behaviour similar to this under vertical collapse, except that the harmonic oscillator is replaced by its 'one-sided' cousion, Airy's operator.

5 b-Fibration Algebra

Returning to the initial setting of a fibration the opposite extreme to the semiclassical case corresponds to trivial fibration in the sense that ϕ has one fibre. Then the operators are simply smooth in the parameter t. This serves to emphasize that the map to $[0, 1]$ is itself a fibration and this can be generalized – as in the setting of the Atiyah-Singer index theorem – to the case of a more general base and fibration with $X \times [0, 1]$ replace by a fibration $\phi : \hat{X} \longrightarrow B$ with typical fibre X. This in turn can be generalized to the case of a *b-fibration* which allows degeneration, of a specific type, on the fibres. Rather than define this in general let me point to a specific type of example.

Suppose M is a compact manifold with corners. It may have many boundary hypersurfaces but each has (by assumption) a defining function – a smooth non-negative function on M which vanishes precisely at the boundary hypersurface in question and has non-zero differential there. Then a total boundary defining function on M is a product of such functions. More generally one can take a product of positive integral powers of such functions. The resulting function, which vanishes at every boundary point but is positive elsewhere, defines a b-fibration – a kind of collar decomposition – as a map to $[0, \epsilon]$ for $\epsilon > 0$ small enough. The general case of a b-fibration is locally the product of such maps and an ordinary fibration. In any case there is a similar structure to the case of a fibration if vector fields tangent to all boundaries are considered and there is an algebra of pseudodifferential operators reducing to the fibre pseudodifferential operators for a fibration:

$$\mathcal{V}_{b/\phi}(M) \subset \rho^\alpha \operatorname{Diff}^*_{b/\phi}(M; W) \subset \rho^\alpha \Psi^*_{b/\phi}(M; W) \subset \rho^\alpha \Psi^{**}_{b/\phi}(M; W) \quad (7)$$

The last space here actually depends on a choice of the resolution of the fibre diagonal ([3]).

Example of this is provided by the *blow-up calculus* being developed with Pierre Albin and also the *gluing calculus* with Michael Singer which corresponds to gluing problems such as treated by Arezzo and Pacard [1]. Namely, from a manifold with an interior separating hypersurface a manifold with corners can be constructed with b-fibration which corresponds to the process of gluing a complete metric on one side of the hypersurface to an appropriate (often incomplete) metric on the other side. The b-fibration is of the type discussed above. This is also related to older work with Andrew Hassell and Rafe Mazzeo on the eta invariant.

6 Morse Degeneration

The two types of calculus above, corresponding to a b-fibration, where the vector fields degenerate only at a submanifold in the boundary, and the adiabatic case where the degenerate to be tangent to the fibres of a fibration can be combined. Rather than set this up in general – corresponding to iterated b-fibrations where there are finer fibrations over (some of) the boundary hypersurfaces of the first b-fibration – let me simply indicate an example which arises from a question of Atiyah.

Every compact manifold M carries a Morse function $f : M \longrightarrow [a, b]$. This can be thought of as a generalization of a fibration over the circle – the setting considered by Witten in [4]. There are singular fibres but they are isolated and of 'minimal singularity'. In particular the singular points, where the differential of f vanishes, are themselves isolated. To construct an adiabatic limit of this b-type, first replace M by the manifold with boundary in which the critical points of f are blown up radially, $M_{\mathrm{Cp}} = [M; \mathrm{Cp}]$ to which f lifts as a smooth function. The singular fibres of f are resolved in the sense that they are each the union of a boundary hypersurface and an embedded (generally non-connected) submanifold SF which meets this boundary transversally. The full space with b-fibration we consider is

$$[M_{\mathrm{Cp}} \times [0, 1]_t; \mathrm{SF}] \longrightarrow [0, 1]_t. \tag{8}$$

The additional condition imposes on vector fields (and hence differential operators) corresponding to the adiabatic limit is that over $t = 0$ they should be tangent to the boundaries, to the regular fibres of f and to the fibres of the blow-down map for the blow-up of SF.

Atiyah's question is whether for a Dirac operator on the total space one can find a formula for the index (which of course is known) in terms of the spectral flow of the induced Dirac operators on the fibres, between the singular values (likely regularized in some way) with perhaps some 'jump terms' across the singular fibres. For the moment I only know how do do this *after* perturbing the operator by a smoothing operator associated to the calculus that I have implicitly described above. To give a more realistic answer requires a better understanding of the behaviour of the eta invariant.

References

1. Claudio Arezzo and Frank Pacard, *On the Kähler classes of constant scalar curvature metrics on blow ups*, arXiv:0706.1838.
2. Joel Fine, *Constant scalar curvature kahler metrics on fibred complex surfaces*, arXiv:0401275.
3. Chris Kottke and Richard Melrose, *Generalized blow-up of corners and fiber products*, Preprint April 2011.
4. E. Witten, *Global gravitational anomalies*, Commun. Math. Phys. **100** (1985), 197–229.

Microlocal Analysis and Adiabatic Problems: The Case of Perturbed Periodic Schrödinger Operators

Gianluca Panati

Microlocal analysis is a powerful technique to deal with multiscale and adiabatic problems in Quantum Mechanics. We illustrate this general claim in the specific case of a perturbed periodic Schrödinger operator, namely the operator defined in a dense subspace of $L^2(\mathbb{R}^d)$ by

$$H_\varepsilon = \frac{1}{2} \sum_{j=1}^{d} \left(-i\frac{\partial}{\partial x_j} - A_j(\varepsilon x) \right)^2 + V(x) + \varphi(\varepsilon x), \tag{1}$$

where $V : \mathbb{R}^d \to \mathbb{R}$ is a $\mathbb{Z}^d$-periodic function, $V \in L^2_{\mathrm{loc}}(\mathbb{R}^d)$, corresponding to the interaction of the test electron with the ionic cores of a crystal, while $A_j \in C^\infty_{\mathrm{b}}(\mathbb{R}^d)$ and $\varphi \in C^\infty_{\mathrm{b}}(\mathbb{R}^d)$ represent some perturbing external electromagnetic potentials. The parameter $\varepsilon \ll 1$ corresponds to the separation of space-scales.

Since the unperturbed Hamiltonian $H_{\mathrm{per}} = -\frac{1}{2}\Delta + V$ is periodic, it can be decomposed as a direct integral of simpler operators, thus exhibiting a *band structure*, analogous to the one appearing in the Born-Oppenheimer problem.

We are interested to the behavior of the solutions to the dynamical Schrödinger equation $i\varepsilon\, \partial_t \psi_\varepsilon(t) = H_\varepsilon \psi_\varepsilon(t)$ in the limit $\varepsilon \to 0$. We show that by using microlocal analysis with operator-valued symbols one can decouple the dynamics corresponding to different bands and determine a simpler approximate dynamics for each band [3]. Further developments have been obtained, more recently, in [1,5].

The Bloch-Floquet transform. The $\mathbb{Z}^d$-symmetry of the unperturbed Hamiltonian operator $H_{\mathrm{per}} = -\frac{1}{2}\Delta + V$ can be used to decomposed it as a direct integral of simpler operators. To fix the notation, let Y be a fundamental domain for the action of the translation group $\Gamma = \mathbb{Z}^d$ on $\mathbb{R}^d$, and let $\mathbb{B}$ be a fundamental domain for the action of the dual lattice $\Gamma^* := \{\kappa \in (\mathbb{R}^d)^* : \kappa \cdot \gamma \in 2\pi\mathbb{Z} \quad \forall \gamma \in \Gamma\}$ on the dual

G. Panati (✉)
Dipartimento di Matematica, Universitá di Roma "La Sapienza", Roma, Italy
e-mail: panati@mat.uniroma1.it

D. Grieser et al. (eds.), *Microlocal Methods in Mathematical Physics and Global Analysis*,
Trends in Mathematics, DOI 10.1007/978-3-0348-0466-0_10, © Springer Basel 2013

space $(\mathbb{R}^d)^*$ ("momentum space"). We also introduce the tori $\mathbb{T}_Y^d = \mathbb{R}^d/\Gamma$ and $\mathbb{T}^* = (\mathbb{R}^d)^*/\Gamma^*$. The formula

$$(\widetilde{\mathcal{U}}\psi)(k, y) = \sum_{\gamma \in \Gamma} e^{-ik\cdot(y+\gamma)}\,\psi(y+\gamma), \qquad y \in \mathbb{R}^d, k \in (\mathbb{R}^d)^*, \psi \in \mathcal{S}(\mathbb{R}^d)$$

extends to a unitary operator $\widetilde{\mathcal{U}} : L^2(\mathbb{R}^d) \longrightarrow L^2(\mathbb{B}) \otimes L^2(\mathbb{T}_Y^d) \simeq L^2(\mathbb{B}, L^2(\mathbb{T}_Y^d))$, called (modified) Bloch-Floquet transform. Hereafter $\mathcal{H}_f := L^2(\mathbb{T}_Y^d)$.

The advantage of this construction is that, after conjugation, H_{per} becomes a fibered operator, namely

$$\widetilde{H}_{per} := \widetilde{\mathcal{U}}\, H_{per}\,\widetilde{\mathcal{U}}^{-1} = \int_{\mathbb{B}}^{\oplus} H_{per}(k)\,dk \ \text{ in } L^2(\mathbb{B}, \mathcal{H}_f) \simeq \int_{\mathbb{B}}^{\oplus} \mathcal{H}_f\,dk =: \mathcal{H},$$

$$H_{per}(k) = \frac{1}{2}(-i\nabla_y + k)^2 + V(y) \ \text{ acting on } \mathcal{D} \subseteq L^2(\mathbb{T}_Y^d, dy) = \mathcal{H}_f$$

where $\mathcal{D}$ is a dense subspace of $\mathcal{H}_f$. The operator $H_{per}(k)$ has compact resolvent, and we label its eigenvalues as $E_0(k) \leq E_1(k) \leq \dots$. Notice that the eigenvalues are Γ^*-periodic. We assume that a solution of the eigenvalue problem $H_{per}(k)\chi_n(k, y) = E_n(k)\chi_n(k, y)$ is known, and we denote by $P_n(k)$ the eigenprojector corresponding to the n-th eigenvalue, while $P_n = \int_{\mathbb{B}}^{\oplus} P_n(k)\,dk$. The set $\mathcal{E}_n = \{(k, E_n(k)) \in \mathbb{T}^* \times \mathbb{R}\}$ is called the n-th Bloch band.

The perturbed dynamics. We consider a Bloch band $\mathcal{E}_n$ which is separated by a gap from the rest of the spectrum, i. e.

$$\inf\{|E_n(k) - E_m(k)| : k \in \mathbb{T}^*, m \neq n\} > 0, \tag{2}$$

and the corresponding subspace

$$\text{Ran}\,P_n = \{\Psi \in \mathcal{H} : \Psi(k, y) = \varphi(k)\,\chi_n(k, y) \text{ for } \varphi \in L^2(\mathbb{B}, dk)\}.$$

In the unperturbed case, $A = 0$ and $\phi = 0$, the subspace $\text{Ran}\,P_n$ is exactly invariant, in the sense that $(1 - P_n)\,e^{-i\widetilde{H}_{pert}/\varepsilon}\,P_n\Psi = 0$ for all $\Psi \in \mathcal{H}$. Moreover, the dynamics of $\Psi \in \text{Ran}\,P_n$ is particularly simple, namely

$$\left(e^{-i\widetilde{H}_{pert}t/\varepsilon}\Psi\right)(k, y) = \left(e^{-iE_n(k)t/\varepsilon}\varphi(k)\right)\chi_n(k, y).$$

Thus a natural question arises: to what extent such properties survive in the perturbed case? More precisely,

(i) Does exist a subspace of $\mathcal{H}$ which is almost-invariant with respect to the dynamics, up to errors of order ε^N?

(ii) Is there any simple (and numerically convenient) way to approximately describe the dynamics inside the almost invariant subspace?

The microlocal approach. Microlocal analysis is a useful tool to answer these questions. In a nutshell, one checks that by modified BF transform one has

$$\widetilde{H}_\varepsilon := \widetilde{\mathcal{U}} \, H_\varepsilon \, \widetilde{\mathcal{U}}^{-1} = \left(-i\nabla_y + k - A(i\varepsilon\nabla_k)\right)^2 + V(y) + \phi(i\varepsilon\nabla_y).$$

The latter operator "looks like" the ε-Weyl quantization of an operator-valued symbol

$$h : \mathbb{T}^* \times \mathbb{R}^d \longrightarrow \text{Operators}(\mathcal{H}_\mathrm{f})$$

$$(k, r) \longmapsto \left(-i\nabla_y + k - A(r)\right)^2 + V(y) + \phi(r).$$

This observation naturally leads to exploit techniques related to matrix-valued pseudo-differential operators [2, 4]. Obviously, to perform this program one has to circumvent some technical *scholia* (*unbounded*-operator-valued symbols, covariance, . . .), for whose solution we refer to [3]. As an answer to question (i), we have the following

Theorem 1. *Let $\mathcal{E}_n$ be an isolated Bloch band, see (2). Then there exists an orthogonal projection $\Pi_{n,\varepsilon} \in \mathcal{B}(\mathcal{H})$ such that for every $N \in \mathbb{N}$ there exist C_N such that*

$$\left\| [\widetilde{H}_\varepsilon, \Pi_{n,\varepsilon}] \right\|_{\mathcal{B}(\mathcal{H})} \leq C_N \, \varepsilon^N$$

and $\Pi_{n,\varepsilon}$ is $\mathcal{O}(\varepsilon^\infty)$-close to the ε-Weyl quantization of a symbol with principal part $\pi_0(k, r) = P_n(k - A(r))$.

As for question (ii), one preliminarily notices that there is no natural identification between $\text{Ran} \Pi_{n,\varepsilon}$ and $L^2(\mathbb{T}^*, dk)$, so no evident reduction of the number of degrees of freedom. To circumvent this obstacle, one constructs an intertwining unitary operator (which is an additional unknown in the problem) $U_{n,\varepsilon} : \text{Ran} \Pi_{n,\varepsilon} \rightarrow L^2(\mathbb{T}^*, dk)$. The freedom to choose $U_{n,\varepsilon}$ can be exploited to obtain a simple and physically transparent representation of the dynamics, as in the following result [3].

Theorem 2. *Let $\mathcal{E}_n$ be an isolated Bloch band. Define the effective Hamiltonian as the operator $\hat{H}_{\mathrm{eff},"} := U_{n,\varepsilon} \, \Pi_{n,\varepsilon} \, H_\varepsilon \, \Pi_{n,\varepsilon} \, U_{n,\varepsilon}^{-1}$ acting in $L^2(\mathbb{T}^*, dk)$. Then:*

(i) (approximation of the dynamics) *for any $N \in \mathbb{N}$ there is C_N such that*

$$\left\| \left(\varepsilon^{-i\widetilde{H}_\varepsilon t/\varepsilon} - U_{n,\varepsilon}^{-1} \, \varepsilon^{-i\hat{H}_{\mathrm{eff},"} \, t/\varepsilon} \, U_{n,\varepsilon} \right) \Pi_{n,\varepsilon} \right\|_{\mathcal{B}(\mathcal{H})} \leq C_N \, \varepsilon^N \, (1 + |t|).$$

(ii) (explicit description of the approximated dynamics) *the operator $\hat{H}_{\mathrm{eff},"}$ is $\mathcal{O}(\varepsilon^\infty)$-close to the ε-Weyl quantization of the symbol $h_\varepsilon^{\mathrm{eff}} : \mathbb{T}^* \times \mathbb{R}^d \rightarrow \mathbb{C}$, with leading orders*

$$h_0^{\mathrm{eff}}(k, r) = E_n(k - A(r)) + \phi(r)$$

$$h_1^{\mathrm{eff}}(k, r) = (\nabla\phi(r) - \nabla E_n(\kappa) \wedge B(r)) + \mathcal{A}_n(\kappa) - B(r) \cdot M_n(\kappa)$$

where $\kappa(k,r) = k - A(r)$, $B_{jl} = \partial_j A_l - \partial_l A_j$, $\mathcal{A}_n(k) = i \langle \chi_n(k) \mid \nabla \chi_n(k) \rangle_{\mathcal{H}_f}$ *is called* Berry connection *and*

$$M_n(k) = \frac{i}{2} \langle \nabla \chi_n(k) \wedge \mid (H_{\mathrm{per}}(k) - E_n(k)) \nabla \chi_n(k) \rangle_{\mathcal{H}_f}.$$

References

1. G. De Nittis, M. Lein. *Applications of magnetic Ψ-DO techniques to Space-adiabatic Perturbation Theory*, Rev. Math. Phys. **23**, 233–260 (2011).
2. C. Emmrich, A. Weinstein. *Geometry of the transport equation in multicomponent WKB approximations*, Commun. Math. Phys. **176**, 701–711 (1996).
3. G. Panati, H. Spohn, S. Teufel. *Effective dynamics for Bloch eelctrons: Peierls substitution and beyond*, Commun. Math. Phys. **242**, 547–578 (2003).
4. J. Sjöstrand. *Projecteurs adiabatiques du point de vue pseudodifferentiel*, C. R. Acad. Sci. Paris Ser. I Math. **317**, 217–220 (1993).
5. H. Stiepan, S. Teufel. Paper in preparation (2011).

Recent Results in Semiclassical Approximation with Rough Potentials

T. Paul

Quantum Mechanics was invented for stability reasons. In fact it is striking to notice the difference of regularity that needs the potential of a Schrödinger operator to insure unitary of the quantum flow (e.g. $V \in L^1_{\text{loc}}$, $\lim_{\epsilon \to 0} \sup_x \int_{|x-y| \le \epsilon} |x - y|^{2-N} |V(y)| dy = 0$) compared to the classical Cauchy-Lipshitz condition for vector fields.

On the other side, tremendous progress have been done in the last 25 years concerning the theorey of ODEs using PDE's methods: extension of the Cauchy-Lipshitz condition to Sobolev ones [5] and BV vector fields (Bouchut for the Hamiltonian case [4] and Ambrosio for the general case [1]) have been proved to provide well-posedness of the classical flow almost everywhere, through uniqness result for the corresponding Liouville equation in the space $L^\infty_+([0, T]; L^1(\mathbb{R}^{2n}) \cap L^\infty(\mathbb{R}^{2n}))$. Under these regularity conditions on the potential (in addition to some growing at infinity) the Schrödinger equation is well posed for all positive values of the Planck constant and it is therefore natural to ask what's happen at the classical limit. As we will see different answers will be given, according to the choice we make first on the topology of the convergence, and secondly on the asymptotic properties of the initial datum. The genral idea of the results we are going to present here can be summarized as follows:

For some $V \notin C^{1,1}$ both the quantum and the classical exist and

the diPerna-Lions-Ambrosio flow is the classical limit of the quantum flow

for non concentrating initial data.

For concentrating initial data

the multivalued bicharateristics are the classical limit of the quantum flow.

T. Paul (✉)
CNRS and CMLS, Ecole Polytechnique, Palaiseau cedex, France
e-mail: thierry.paul@math.polytechnique.fr

D. Grieser et al. (eds.), *Microlocal Methods in Mathematical Physics and Global Analysis*,   49
Trends in Mathematics, DOI 10.1007/978-3-0348-0466-0_11, © Springer Basel 2013

All the results presented here will use a quantum formalsim on phase space, thanks to the notion of Wigner funtion. More precisely we will be concerned with the so-called Schrödinger and von Neumann equation

$$i\hbar\partial_t\psi = (-\hbar^2\Delta + V)\psi \text{ and } \partial_t D = \frac{1}{i\hbar}[-\hbar^2\Delta + V, D]$$

with $\psi^{t=0} \in L^2(\mathbb{R}^n)$ and $D^{t=0} \geq 0, TrD^{t=0} = 1$ (density matrix, e.g. $D^0 = |\psi^0\rangle\langle\psi^0|$). And we will consider the Wigner funtion associced to D^t (e.g. $= |\psi^t\rangle\langle\psi^t|$), defined by

$$W^\epsilon D(x, p) := \frac{1}{(2\pi)^n} \int_{\mathbb{R}^n} D^t(x + \frac{\epsilon}{2}y, x - \frac{\epsilon}{2}y)e^{-ipy}dy$$

where $D^t(x, y)$ is the integral kernel of D^t (e.g. $= \overline{\psi^t(x)}\psi^t(y)$ in which case we write $W^\epsilon\psi^t_\epsilon$).

The well-known lack of positivity of W^ϵ suggests, in order to study evolution in spaces like L^∞_+, to use the so-called Husimi function of D^t, a molification of W^ϵ defined as $\widetilde{W^\epsilon D} := e^{\epsilon\Delta_{\mathbb{R}^{2n}}}W^\epsilon D$ which happens to be positive. But the only bound we have for $\widetilde{W^\epsilon D}$ is $\|\widetilde{W^\epsilon D}\|_{L^\infty} \leq \epsilon^{-n}TrD$, unuseful for the L^∞ condition needed for the existence of the classical solution. We formulate the

Conjecture. For an ϵ dependant family D_ϵ of density matrices we have

$$TrD_\epsilon = 1 \implies \sup_{\epsilon>0}\|\widetilde{W^\epsilon D_\epsilon}\|_{L^\infty} = +\infty$$

In the general case of a potentials whose gradient is BV, the first idea will be to smeared out the initial conditions and consider a family of vectors $\psi^\epsilon_{\epsilon,w}$, w belonging to a probabilty space $(W, \mathcal{F}, \mathbb{P})$. Under the general assumptions

Assumptions on V	Assumptions on initial datum
globally bounded, locally Lipschitz	$\psi^\epsilon_{0,w} \in H^2(\mathbb{R}^n; \mathbb{C})$
$\nabla U_b \in BV_{loc}(\mathbb{R}^n; \mathbb{R}^n)$	$\sup_{\epsilon>0}\int_W\int_{\mathbb{R}^n}|H_\epsilon\psi^0_{\epsilon,w}|^2dxd\mathbb{P}(w)<\infty$
ess $\sup_{x\in\mathbb{R}^n}\frac{|\nabla U_b(x)|}{1+|x|} < +\infty$	$\int_W|\psi^0_{\epsilon,w}\rangle\langle\psi^0_{\epsilon,w}|\,d\mathbb{P}(w) \leq \epsilon^n\text{Id}$
+ finite repulsive Coulomb singularities	$\lim_{\epsilon\downarrow 0}\widetilde{W^\epsilon\psi^0_{\epsilon,w}} = i(w) \in \mathcal{P}(\mathbb{R}^d)$
	for $\mathbb{P} - a.e.\ w \in W$.

we have, any bounded distance $d_\mathcal{P}$ inducing the weak topology in $\mathcal{P}(\mathbb{R}^{2n})$, the

Theorem 1 ([2]).

$$\lim_{\epsilon\to 0}\int_W \sup_{t\in[-T,T]} d_\mathcal{P}(\widetilde{W^\epsilon\psi^t_{\epsilon,w}}), \mu(t, i(w)))\,d\mathbb{P}(w) = 0,$$

where $\mu(t, v)$ is a (regular Lagrangian) flow on $\mathcal{P}(\mathcal{P}(\mathbb{R}^{2n}))$ "solving" the Liouville equation.

In the case of the von Neumann equation, a more direct result can be obtained.

Assumptions on V Assumptions on initial datum

globally bounded, locally Lipschitz $\sup_{\epsilon \in (0,1)} \mathrm{Tr}(H_\epsilon^2 D_\epsilon^o) < +\infty$

$\nabla U_b \in BV_{loc}(\mathbb{R}^n; \mathbb{R}^n)$ $D_\epsilon^o \leq \epsilon^n \mathrm{Id}$

$\mathrm{ess}\sup_{x \in \mathbb{R}^n} \frac{|\nabla U_b(x)|}{1+|x|} < +\infty$ $w - \lim_{\epsilon \to 0} W^\epsilon D_\epsilon^0 = W_0^0 \in \mathcal{P}\mathbb{R}^{2n}$

Theorem 2 ([6]). *Let $d_{\mathcal{P}}$ be any bounded distance inducing the weak topology in $\mathcal{P}(\mathbb{R}^{2n})$. Then*

$$\limsup_{\epsilon \to 0 \, [0,T]} d_{\mathcal{P}}(\widetilde{W^\epsilon D_\epsilon^t}, W_t^0) = 0,$$

W_t^0 *is the unique solution in $L_+^\infty([0, T]; L^1(\mathbb{R}^{2n}) \cap L^\infty(\mathbb{R}^{2n}))$ of the Liouville equation.*

The next result concern the semiclassical approximation in strong topology. Let us denote $\widetilde{V} := e^{\epsilon \Delta_{\mathbb{R}^n}} V$.

(new) Assumptions on V (new) Assumptions on initial datum

$\int |\widehat{V}(S)| \frac{|S|^2}{1+|S|^2} \, dS < \infty$ $W_0^\epsilon \in H^2(\mathbb{R}^n)$

$\int\limits_{|S| \in (a,b)} |\widehat{V}(S)| \, |S|^m dS \leq$ *if* $\partial_t \rho + k \partial_x \rho - \partial_x \widetilde{V} \cdot \partial_x \rho = 0$, $\rho_{t=0} := W_0^\epsilon$

$C \left(b^{m-1-\theta} - a^{m-1-\theta} \right)$

$m = 0, 1, 2, \, 0 < \theta < 1$ $\exists T > 0, \delta \in \left(0, \frac{\theta}{2+\theta} \right)$ *such that*

$$\|\rho(t)\|_{H^2} = O(\varepsilon^{-\delta} \|W_0^\varepsilon\|_{L^2}) \text{ for } t \in [0, T]$$

Theorem 3 ([3]). *Let ρ_1^ϵ be the solution of*

$$\partial_t \rho_1^\epsilon + k \partial_x \rho_1^\epsilon - \partial_x \widetilde{V} \cdot \partial_x \rho_1^\epsilon = 0.$$

$W_t^\epsilon := W^\epsilon D_\epsilon^t$ *satisfies, uniformly on $[0, T]$,*

$$\|W_t^\epsilon - \rho_1^\epsilon(t)\|_{L^2} = O(\varepsilon^\kappa \|W_0^\varepsilon\|_{L^2}), \; \kappa = \min\{ \frac{1+\theta}{2} - 1, \frac{\theta}{2+\theta} - \delta \}.$$

Let us now give a 1D example where the lack of unicity will be crucial.

Let V be a confining potentail such that $V = -|x|^{1+\theta}$ near 0. Near $(0, 0)$ we obtain two solutions of the Hamiltonian flow:

$$(X^\pm(t), P^\pm(t)) = (\pm c_0 t^v, \pm c_0 v t^{v-1}),$$

$\nu = \frac{2}{1-\theta}$ and $c_0 = \left(\frac{(1-\theta)^2}{2}\right)^{1-\theta}$, plus a continuum family of solutions by not moving up to any value of the time and then starting to move according to $(X^{\pm}(t), P^{\pm}(t))$.

The question now is to know which one is going to be selected the semiclassical limit. The answer is given by the following result.

Theorem 4 ([3]). *Let* $W^{\varepsilon} D_{\epsilon}^0(x, k) = \lambda^{\frac{7+3\theta}{30}} w(\lambda^{\frac{1+\theta}{6}} x, \lambda^{\frac{1-\theta}{15}} k)$, $\lambda = \log \frac{1}{\epsilon}$, *supp* $w \subseteq$ $\{|x|^2 + |k|^2 < 1\}$.
Then $\exists T > 0/t \in [0, T]$, $W^{\epsilon} D_{\epsilon}^t$ *converges in weak-*$*$ sense to*

$$W_t^0 = c_+ \delta_{(X^+(t), P^+(t))} + c_- \delta_{(X^-(t), P^-(t))},$$

$$c_{\pm} = \int\limits_{\pm x > 0} w(x, k) dx dk.$$

What these results show is the fact that, at the contrary of the case where the underlying classical dynamics is well-posed, the semiclassical limit of the qunatum evolution with non regular (i.e. not providing uniqness of the classical flow) potentials is not unique, and depends on the family itself of initial conditions, and not anymore only on their limit.

For non concentrating data the classical limit, in the general case of a potential whose gradient is BV, is driven (in the two senses expressed by Thoerems 1 and 2) by the DiPerna-Lions-Bouchut-Ambrosio flow.

Slowly concentrating data (Theorem 4) provide situations where the classical limit is ubiquous, and follows several of the non unique bi-charateritics, a typical quantum feature surviving in this situation the classical limit. It is important to remark that the speed of concentration governs the selections of the remaining trajectories. The case of fast concentration, in particular the pure states situations, is still open.

References

1. L. Ambrosio: Transport equation and Cauchy problem for BV vector fields. Invent. Math., 158 (2004), 227-260.
2. L. Ambrosio, A. Figalli, G. Friesecke, J. Giannoulis & T. Paul: Semiclassical limit of quantum dynamics with rough potentials and well posedness of transport equations with measure initial data, Comm. Pure Appl. Math., 64 (2011),1199-1242.
3. A. Athanassoulis & T. Paul: Strong and weak semiclassical limits for some rough Hamiltonians, Mathematical Models and Methods in Applied Sciences, 12 (22) (2012).
4. F. Bouchut: Renormalized solutions to the Vlasov equation with coecients of bounded variation. Arch. Ration. Mech. Anal., 157 (2001), 75-90.
5. R.J. DiPerna, P.L. Lions: Ordinary differential equations, transport theory and Sobolev spaces. Invent. Math., 98 (1989), 511-547.
6. A. Figalli, M. Ligabo & T. Paul: Semiclassical limit for mixed states with singular and rough potentials, to appear in "Indiana University Mathematics Journal".

Part II
Singular spaces

On the Closure of Elliptic Wedge Operators

Juan B. Gil, Thomas Krainer, and Gerardo A. Mendoza

2010 Mathematics Subject Classification: Primary: 58J50; Secondary: 35P05, 58J32, 58J05.

We present a semi-Fredholm theorem for the minimal extension of an elliptic differential operator on a manifold with wedge singularities and give, under suitable assumptions, a full asymptotic expansion of the trace of the resolvent.

1 Wedge Operators

Let $\mathcal{M}$ be a smooth compact manifold with boundary. Assume that the boundary is the total space of a locally trivial fiber bundle $\wp : \partial\mathcal{M} \to \mathcal{Y}$ with typical fiber $\mathcal{Z}$, where $\mathcal{Y}$ and $\mathcal{Z}$ are smooth compact manifolds. Let $E, F \to \mathcal{M}$ be smooth vector bundles. We are interested in the space $x^{-m} \operatorname{Diff}_e^m(\mathcal{M}; E, F)$ of differential *wedge operators* of order m, where $\operatorname{Diff}_e^m(\mathcal{M}; E, F)$ denotes the space of differential edge operators, as introduced in [3], and $x : \mathcal{M} \to \mathbb{R}$ is any smooth defining function for $\partial\mathcal{M}$.

Locally, near a point $p \in \partial\mathcal{M}$, a wedge operator $A \in x^{-m} \operatorname{Diff}_e^m(\mathcal{M}; E, F)$ can be represented as

$$A = x^{-m} \sum_{k+|\alpha|+|\beta|\leq m} a_{k,\alpha,\beta}(x, y, z)(xD_x)^k (xD_y)^\alpha D_z^\beta \tag{1}$$

with coefficients $a_{k,\alpha,\beta}$ smooth up to $x = 0$.

J.B. Gil (✉) · T. Krainer
Penn State Altoona, 3000 Ivyside Park, Altoona, PA, 16601, USA
e-mail: jgil@psu.edu; krainer@psu.edu

G.A. Mendoza
Department of Mathematics, Temple University, Philadelphia, PA 19122, USA
e-mail: gmendoza@temple.edu

D. Grieser et al. (eds.), *Microlocal Methods in Mathematical Physics and Global Analysis*,   55
Trends in Mathematics, DOI 10.1007/978-3-0348-0466-0_12, © Springer Basel 2013

For example, if g_w is a Riemannian metric on $\mathcal{M}$ that near $\partial\mathcal{M}$ takes the form $g_w = dx^2 + x^2 g_{\mathcal{Z}} + g_{\mathcal{Y}}$ (wedge metric), then the Laplacian associated with g_w is a wedge operator of order 2 and has the local representation

$$x^{-2}\big((xD_x)^2 - i(\dim \mathcal{Z} - 1)(xD_x) + \Delta_{\mathcal{Z}} + x^2\Delta_{\mathcal{Y}}\big).$$

Note that if $\mathcal{Y} = \{\text{pt}\}$, the space $x^{-m}\operatorname{Diff}_e^m(\mathcal{M}; E, F)$ reduces to the class of cone operators, and if $\mathcal{Z} = \{\text{pt}\}$, then $x^{-m}\operatorname{Diff}_e^m(\mathcal{M}; E, F)$ contains all regular differential operators on $\mathcal{M}$.

Let A be a wedge operator or order m, locally near the boundary represented as in (1). The following principal symbols are intrinsically associated with A.

The w-principal symbol. There is a natural structure bundle $^wT^*\mathcal{M} \to \mathcal{M}$ associated with wedge geometry. The principal symbol of a wedge operator A extends from the interior of $\mathcal{M}$ to $^wT^*\mathcal{M} \setminus 0$. Locally near $\partial\mathcal{M}$, the w-principal symbol of A can be represented as

$$^w\boldsymbol{\sigma}(A) = \sum_{k+|\alpha|+|\beta|=m} a_{k,\alpha,\beta}(x, y, z)\xi^k \eta^\alpha \zeta^\beta.$$

The operator A is said to be w-elliptic if $^w\boldsymbol{\sigma}(A)$ is invertible.

The conormal symbol (indicial family). The indicial family restricts to the fibers of $\wp : \partial\mathcal{M} \to \mathcal{Y}$. We have

$$\hat{A}(y, \sigma) = \sum_{k+|\beta|\leq m} a_{k,0,\beta}(0, y, z)\sigma^k D_z^\beta$$

for $y \in \mathcal{Y}$ and $\sigma \in \mathbb{C}$, and $\hat{A}(y, \sigma)$ acts on $C^\infty(\mathcal{Z})$. The set

$$\operatorname{spec}_e(A) = \{(y, \sigma) \in \mathcal{Y} \times \mathbb{C} : \hat{A}(y, \sigma) \text{ is not invertible}\}$$

is called the boundary spectrum of A.

The principal edge symbol (normal family). The choice of defining function trivializes the inward pointing half $\mathcal{N}_+(\partial\mathcal{M})$ of the normal bundle of $\partial\mathcal{M}$ in $\mathcal{M}$. We get an induced fibration $\wp_\wedge : \mathcal{N}_+(\partial\mathcal{M}) \to \mathcal{Y}$ with typical fiber $\mathcal{Z}^\wedge = \overline{\mathbb{R}}_+ \times \mathcal{Z}$. Locally, the normal family takes the form

$$A_\wedge(y, \eta) = x^{-m} \sum_{k+|\alpha|+|\beta|\leq m} a_{k,\alpha,\beta}(0, y, z)(xD_x)^k (x\eta)^\alpha D_z^\beta,$$

where $(y, \eta) \in T^*\mathcal{Y} \setminus 0$, and $A_\wedge(y, \eta)$ acts in the canonically induced conic L^2-space on the fiber $\mathcal{Z}^\wedge$.

Under suitable conditions on the normal family and the boundary spectrum of a w-elliptic wedge operator, we present the following results, cf. [2].

2 Main Results

For simplicity of the exposition, we assume that the operators are scalar.

Let $A \in x^{-m} \operatorname{Diff}_e^m(\mathcal{M})$ be w-elliptic, considered as an unbounded operator

$$A : C_c^\infty(\overset{\circ}{\mathcal{M}}) \subset x^{-\gamma} L_b^2(\mathcal{M}) \to x^{-\gamma} L_b^2(\mathcal{M})$$

for some fixed $\gamma \in \mathbb{R}$. Here $L_b^2(\mathcal{M})$ denotes the L^2 space defined using a fixed density of the form $x^{-1}\mathfrak{m}$ for some smooth positive density $\mathfrak{m}$ on $\mathcal{M}$.

We let $H_e^m(\mathcal{M})$ denote the corresponding Sobolev space defined using edge differential operators of order $\leq m$. Let $\mathcal{D}_{\min}$ be the closure of $C_c^\infty(\overset{\circ}{\mathcal{M}})$ with respect to the graph norm of A, and let $\mathcal{D}_{\wedge,\min}(y)$ be the closure of $C_c^\infty(\mathcal{Z}^\wedge)$ in $x^{-\gamma} L_b^2(\mathcal{Z}^\wedge)$ with respect to the one of $A_\wedge(y, \eta)$.

Our first result concerns the minimal domain and the semi-Fredholm property of the minimal extension of A.

Theorem 1. *Let A be as above. If $A_\wedge(y, \eta) : \mathcal{D}_{\wedge,\min}(y) \to x^{-\gamma} L_b^2$ is injective on $T^*\mathcal{Y} \setminus 0$, and if $\pi_\mathbb{C} \operatorname{spec}_e(A) \cap \{\Im \sigma = \gamma - m\} = \emptyset$, then $\mathcal{D}_{\min}(A) = x^{-\gamma+m} H_e^m(\mathcal{M})$ and $A : \mathcal{D}_{\min} \to x^{-\gamma} L_b^2(\mathcal{M})$ is a semi-Fredholm operator with finite-dimensional kernel and closed range.*

For our next result, let Λ be a closed sector properly contained in $\mathbb{C}$. Such a sector is called a *sector of minimal growth* for $A_\mathcal{D} : \mathcal{D} \subset x^{-\gamma} L_b^2 \to x^{-\gamma} L_b^2$, if $A_\mathcal{D} - \lambda$ is invertible for $|\lambda|$ large, and $\|(A_\mathcal{D} - \lambda)^{-1}\|_{\mathscr{L}(x^{-\gamma} L_b^2)} = O(|\lambda|^{-1})$ as $|\lambda| \to \infty$.

Theorem 2 (Resolvent expansion). *Let $m > 0$, let $A \in x^{-m} \operatorname{Diff}_e^m(\mathcal{M})$ be such that $\operatorname{spec}(\overset{w}{\sigma}(A)) \cap \Lambda = \emptyset$ on $^wT^*\mathcal{M} \setminus 0$. If $\pi_\mathbb{C} \operatorname{spec}_e(A) \cap \{\Im \sigma = \gamma - m\} = \emptyset$, and if $A_\wedge(y, \eta) - \lambda : \mathcal{D}_{\wedge,\min}(y) \to x^{-\gamma} L_b^2$ is bijective on $(T^*\mathcal{Y} \times \Lambda) \setminus 0$, then Λ is a sector of minimal growth for $A_{\mathcal{D}_{\min}}$, and for every $\ell \in \mathbb{N}$ with $\ell > \frac{\dim \mathcal{M}}{m}$,*

$$\left(A_{\mathcal{D}_{\min}} - \lambda\right)^{-\ell} : x^{-\gamma} L_b^2(\mathcal{M}) \to x^{-\gamma} L_b^2(\mathcal{M})$$

is of trace class. For every $\varphi \in C^\infty(\mathcal{M})$, we have an expansion

$$\operatorname{Tr}\left(\varphi\left(A_{\mathcal{D}_{\min}} - \lambda\right)^{-\ell}\right) \sim \sum_{j=0}^\infty \sum_{k=0}^{m_j} \alpha_{jk} \lambda^{\frac{\dim \mathcal{M} - j}{m} - \ell} \log^k(\lambda) \text{ as } |\lambda| \to \infty. \tag{2}$$

Here $m_j \leq 1$ for all j, and $m_j = 0$ for $j \leq \dim \mathcal{Z}$.

By standard methods, this asymptotic expansion leads to short time asymptotics of the heat trace when $A_{\mathcal{D}_{\min}}$ is sectorial, and to results concerning the meromorphic structure of the ζ-function when $A_{\mathcal{D}_{\min}}$ is positive.

The above theorems rely on the construction of suitable parametrices within the class of wedge pseudodifferential operators. Our approach makes substantial use of pseudodifferential methods developed by Schulze, see e.g. [4].

The asymptotic expansion in (2) is of course consistent with what is known in the special cases of boundary value problems (dim $\mathcal{Z} = 0$) and of elliptic cone operators (dim $\mathcal{Y} = 0$). For closed extensions other than the minimal extension, one should generally expect a more intricate asymptotic structure of the resolvent. In fact, in the case when dim $\mathcal{Y} = 0$, the corresponding expansion of the resolvent sometimes involves rational functions in $\log \lambda$ and complex powers of λ, see [1].

Acknowledgements Work partially supported by the NSF, grants DMS-0901173 & DMS-0901202.

References

1. J. Gil, T. Krainer, and G. Mendoza, *Dynamics on Grassmannians and resolvents of cone operators*, Anal. PDE **4** (2011), no. 1, 115–148.
2. J. Gil, T. Krainer, and G. Mendoza, *On the closure of elliptic wedge operators*, preprint arXiv:1007.2397v2, December 2010.
3. R. Mazzeo, *Elliptic theory of differential edge operators I*, Comm. Partial Differential Equations **16** (1991), 1615–1664.
4. B.-W. Schulze, *Pseudo-differential Operators on Manifolds with Singularities*, North Holland, Amsterdam, 1991.

Generalized Blow-Up of Corners and Fiber Products

Chris Kottke and Richard Melrose

Consider the category of manifolds with corners and interior b-maps $f : X \to Y$. These are required to pull back smooth functions to be smooth, and pull back each principal ideal[1] $\mathcal{I}_H = C^\infty(Y) \cdot \rho_H$ of functions vanishing on a boundary hypersurface $H \in \mathcal{M}_1(Y)$ to a product

$$f^* \mathcal{I}_H \subset \prod_{G \in \mathcal{M}_1(X)} \mathcal{I}_G^{\alpha(H,G)} \quad \alpha(\cdot, \cdot) \in \mathbb{N} \tag{1}$$

of similar ideals in $C^\infty(X)$. One reason to consider this category is that it contains blow-up.

Recall that the blow-up of a codimension k boundary face[2] $F \in \mathcal{M}_k(Y)$ is the space $[Y ; F] = Y \setminus F \cup S_+ NF$, where $S_+ NF$ denotes the inward pointing spherical normal bundle. It has a "blow-down" map $\beta : [Y ; F] \to Y$ and is equipped with the smooth functions generated by $\beta^* C^\infty(Y)$ as well as the quotients ρ_{H_i} / ρ_{H_j} (where finite) of boundary defining functions for hypersurfaces through F.

While this theory is well-known, we give a new description of the data defined by a b-map that allows for significant clarification and generalization of boundary blow-up, which we use to discuss the existence and resolution of fiber products of

---

[1] Here ρ_H is a boundary defining function for H – a nonnegative smooth function vanishing simply and exactly on H.

[2] We only consider the blow-up of boundary faces and its subsequent generalization, leaving the situation of general submanifolds to a future work.

C. Kottke (✉)
Brown University, Providence, USA
e-mail: ckottke@math.brown.edu

R. Melrose
Department of Mathematics, MIT, Cambridge, MA, USA
e-mail: rbm@math.mit.edu

D. Grieser et al. (eds.), *Microlocal Methods in Mathematical Physics and Global Analysis*,   59
Trends in Mathematics, DOI 10.1007/978-3-0348-0466-0_13, © Springer Basel 2013

manifolds with corners. This new description is the theory of monoidal complexes and their refinements.

The boundary faces of a manifold have natural 'b-normal' spaces

$$^bNF \subset {}^bT_FX, \quad F \in \mathcal{M}_*(X)$$

with natural inclusions $^bN_pF \subset {}^bN_pG$ whenever $p \in G \subset F$. At each point these are spanned by the 'radial' vector fields with respect to the face in question. As a result such a bundle has a global canonical frame[3] $\{\rho_i \partial_{\rho_i}\}$ by which it can be trivialized, identifying the fibers with a fixed vector space ^bNF which has well-defined lattice structure $\text{span}_{\mathbb{Z}}\{\rho_i \partial_{\rho_i}\}$. Taking the inward pointing lattice points defines a 'smooth,' which is to say freely generated, monoid

$$\sigma_F = \text{span}_{\mathbb{Z}_+}\{\rho_i \partial_{\rho_i}\},$$

and the collection of these along with the inclusions $i_{GF} : \sigma_G \hookrightarrow \sigma_F$ for $G \subseteq F$ define what we call the 'basic monoidal complex' of X:

$$\mathcal{P}_X = \{(\sigma_F, i_{GF}) ; G \subseteq F \in \mathcal{M}(X)\}.$$

A b-map $f : X \to Y$ has a tangent differential which at a face $F \in \mathcal{M}(X)$ restricts to a well-defined monoid homomorphism (i.e. additive map)

$$f_\natural : \sigma_F \to \sigma_G$$

where G is the boundary face of largest codimension in Y such that $f(F) \subset G$. Indeed, viewed as a matrix, the coefficients of this map are just the relevant exponents $\alpha(\cdot, \cdot) \in \mathbb{Z}_+$ in (1). The collection of these homomorphisms patch together to form a morphism

$$f_\natural : \mathcal{P}_X \to \mathcal{P}_Y$$

of monoidal complexes which is fundamental to our discussion.

In general, the monoidal complexes and their morphisms capture only the combinatorial relationships between boundary faces of X, those of Y, and the order of vanishing of boundary defining functions with respect to these faces. However, in the case of blow-up, this is enough to completely specify the domain $X = [Y; F]$ in terms of the range Y. Indeed, in this case the blow-down map has additional properties, namely

$$\beta : X \setminus \partial X \to Y \setminus \partial Y \text{ is a diffeomorphism,} \tag{2}$$

---

[3]The ρ_i are boundary defining functions for the hypersurfaces through F defined in a neighborhood.

and

$$^{b}\beta_* : {}^{b}T_p X \to {}^{b}T_{\beta(p)}Y \text{ is an isomorphism for all } p, \tag{3}$$

and the morphism $\beta_\natural : \mathcal{P}_X \to \mathcal{P}_Y$ forms what we call a 'smooth refinement' of $\mathcal{P}_Y$.

Abstracting this, we call a smooth proper map between manifolds satisfying (1), (2) and (3) a *generalized blow-down map*. Such are substantially more general than standard blow-down maps, and one of our main results is a complete characterization of these maps.

Theorem 1. *A generalized blow-down map* $f : X \to Y$ *determines a smooth refinement* $\mathcal{P}_X \to \mathcal{P}_Y$ *of the monoidal complex on* Y, *and conversely for any smooth refinement* $\mathcal{R} \to \mathcal{P}_Y$ *there is a unique (up to diffeomorphism) manifold* $X = [Y ; \mathcal{R}]$ *with* $\mathcal{P}_X = \mathcal{R}$ *and a generalized blow-down map* $f : X = [Y ; \mathcal{R}] \to Y$.

We call $[Y ; \mathcal{R}]$ the 'generalized blow-up' of Y by the refinement $\mathcal{R}$, and we show that the important question of lifting of b-maps under generalized blow-ups of the domain and/or range can be addressed at the level of monoidal complexes.

Finally this theory is applied to the problem of fiber products. Recall that, in any category, the fiber product of two maps $f_i : X_i \to Y$, $i = 1, 2$ is an object X with maps $h_i : X \to X_i$ such that $f_1 \circ h_1 = f_2 \circ h_2$, and has the *universal property* that for any other maps $g_i : Z \to X_i$ such that $g_2 \circ f_2 = g_1 \circ f_1$ there is a unique map $h : Z \to X$ through which they factor.

In the category of sets there is a unique fiber product

$$X_1 \times_Y X_2 = \{(p_1, p_2) ; \ f_1(p_1) = f_2(p_2)\} \subset X_1 \times X_2, \tag{4}$$

however, in the setting of manifolds, (4) is not smooth and fiber products do not generally exist. For manifolds without boundary, there is a well-known sufficient condition for existence, namely that f_1 and f_2 be *transversal*, meaning that whenever $f_1(p_1) = f_2(p_2) = q \in Y$, then

$$(f_1)_*(T_{p_1} X_1) + (f_2)_*(T_{p_2} X_2) = T_q Y. \tag{5}$$

In this case (4) is a smooth manifold and the h_i are smooth maps.

The natural analog of (5) in the setting of manifolds with corners is 'b-transversality,' namely the requirement that

$$(^{b}f_1)_*(^{b}T_{p_1} X_1) + (^{b}f_2)_*(^{b}T_{p_2} X_2) = {}^{b}T_q Y. \tag{6}$$

Under this condition, (4) is not necessarily a manifold, but it is a union of what we call 'interior binomial subvarieties.' These are objects generalizing manifolds with corners, with smooth interiors and boundary faces of the same type.

As for a manifold, there is a natural monoidal complex $\mathcal{P}_D$ defined over the boundary faces of a binomial subvariety $D \subset X$, the difference being that the monoids may not be smooth (freely generated). If they *are* smooth, then D has a natural structure of a smooth manifold (though it need not be smoothly embedded in X), and if they are not, we show that D can be resolved, giving a smooth manifold $[D; \mathcal{R}] \to D$ for every smooth refinement $\mathcal{R} \to \mathcal{P}_D$.

In the case of fiber products, the monoids in $\mathcal{P}_D$ are of the form

$$\sigma_{F_1} \times_{\sigma_G} \sigma_{F_2}, \quad F_i \in \mathcal{M}(X_i), \ f_i(F_i) \subset G \tag{7}$$

which leads to our second main theorem.

Theorem 2. *If $f_i : X_i \to Y$ are b-maps of manifolds with corners which satisfy* (6), *and if each of the monoids* (7) *is freely generated, then there exists a smooth fiber product in the category of manifolds with corners.*

In case the monoids (7) are not freely generated, our theory leads to the following 'resolved' version of the fiber product.

Theorem 3. *For every smooth refinement $\mathcal{R}$ of the complex $\mathcal{P}_{X_1 \times_Y X_2}$, there is a smooth manifold with corners $[X_1 \times_Y X_2 ; \mathcal{R}]$ with maps to X_i commuting with the $f_i : X_i \to Y$. If $h_i : Z \to X_i$, $i = 1, 2$ are smooth maps commuting with the f_i for some other manifold Z, then there exists a generalized blow-up $[Z; \mathcal{S}] \to Z$ and a unique map $h : [Z; \mathcal{S}] \to [X_1 \times_Y X_2 ; \mathcal{R}]$ such that the maps form a commutative diagram.*

Trace Expansions for Elliptic Cone Operators

Thomas Krainer, Juan B. Gil, and Gerardo A. Mendoza

2010 Mathematics Subject Classification: Primary: 58J35; Secondary: 35P05, 47A10.

1 Introduction

Let (M, g) be a Riemannian manifold and $E \to M$ a Hermitian vector bundle. Let

$$A : C_c^\infty(M; E) \to C^\infty(M; E) \tag{1}$$

be a differential operator (with smooth coefficients). A is closable as an unbounded operator in $H = L^2(M; E)$. The domains of the minimal and maximal extension of A are

$\mathcal{D}_{\min} = $ Closure of (1) with respect to the graph norm $\|u\|_A = \|u\|_{L^2} + \|Au\|_{L^2}$,

$\mathcal{D}_{\max} = \{u \in L^2(M; E);\ Au \in L^2(M; E)\}.$

Both are complete in the graph norm, and the closed extensions of A are the unbounded operators $A_{\mathcal{D}}$ in H that act like A with domains $\mathcal{D}_{\min} \subset \mathcal{D} \subset \mathcal{D}_{\max}$ such that $\mathcal{D}$ is complete in the graph norm.

T. Krainer (✉) · J.B. Gil
Penn State Altoona, 3000 Ivyside Park, Altoona, PA, 16601-3760, USA
e-mail: krainer@psu.edu; jgil@psu.edu

G.A. Mendoza
Department of Mathematics, Temple University, Philadelphia, PA 19122, USA
e-mail: gmendoza@temple.edu

D. Grieser et al. (eds.), *Microlocal Methods in Mathematical Physics and Global Analysis*, 63
Trends in Mathematics, DOI 10.1007/978-3-0348-0466-0_14, © Springer Basel 2013

Let $\Lambda = \{\lambda = re^{i\theta}; \ r \geq 0, \ |\theta - \theta_0| \leq \alpha\}$ be a closed sector. Suppose that Λ is a sector of minimal growth for $A_\mathcal{D}$, i.e., $A - \lambda : \mathcal{D} \to H$ is invertible for all $\lambda \in \Lambda$ with $|\lambda| > 0$ sufficiently large, and

$$\|(A_\mathcal{D} - \lambda)^{-1}\|_{\mathcal{L}(H)} = O(|\lambda|^{-1}) \text{ as } |\lambda| \to \infty, \ \lambda \in \Lambda.$$

Furthermore, assume that $(A_\mathcal{D} - \lambda_0)^{-1} \in C_p(H)$ for some $\lambda_0 \in \mathbb{C}$ and $p > 0$, where $C_p(H)$ denotes the Schatten class of all operators $T \in \mathcal{L}(H)$ such that the sequence of nonzero eigenvalues of $\sqrt{T^*T}$ (counting multiplicities) is p-summable.

Elementary functional analysis then shows that, for $\ell \in \mathbb{N}$ large enough, the ℓ-th power of the resolvent $(A_\mathcal{D} - \lambda)^{-\ell} : H \to H$ is of trace class, and the trace

$$t(\lambda) = \text{Tr}(A_\mathcal{D} - \lambda)^{-\ell} : \Lambda_R \to \mathbb{C} \tag{2}$$

is a symbol on $\Lambda_R = \{\lambda \in \Lambda; \ |\lambda| > R\}$ for $R > 0$ large enough. We will refer to (2) as the *resolvent trace*.

The following question is motivated by Seeley's seminal work [19] on complex powers of elliptic operators.

Question 1. What are the asymptotics of the resolvent trace $\text{Tr}(A_\mathcal{D} - \lambda)^{-\ell}$ on Λ as $|\lambda| \to \infty$?

The asymptotic properties of the resolvent trace immediately translate to structural results about the short time asymptotics of the heat trace $\text{Tr} \, e^{-tA_\mathcal{D}}$ if $A_\mathcal{D}$ is sectorial, and to results about the singularities of the analytic extension of the ζ-function associated with a semibounded operator $A_\mathcal{D}$. These in turn have numerous applications in global analysis.

If M is a closed, compact manifold of dimension n, and $A \in \text{Diff}^m(M; E)$, $m > 0$, is elliptic with parameter in Λ, i.e., the principal symbol $\varpi(A)$ has no spectrum in Λ everywhere on $T^*M \setminus 0$, then Seeley's analysis shows that for $\ell > n/m$ the resolvent trace (2) is a step-$\frac{1}{m}$ polyhomogeneous symbol of order $\frac{n-\ell m}{m}$ on Λ_R, i.e.,

$$\text{Tr}(A - \lambda)^{-\ell} \sim \sum_{j=0}^{\infty} \alpha_j \lambda^{\frac{n-\ell m-j}{m}} \text{ as } |\lambda| \to \infty. \tag{3}$$

Note that $A : C^\infty(M; E) \to C^\infty(M; E)$ is essentially closed in $L^2(M; E)$ with $\mathcal{D}_{\min} = \mathcal{D}_{\max} = H^m(M; E)$ in this case. The same asymptotics also hold for the resolvent trace of L^2-realizations of parameter-dependent elliptic operators A on smooth compact manifolds $\overline{M}$ with boundary subject to differential boundary conditions $Tu = 0$ on $\partial\overline{M}$ that satisfy the Shapiro-Lopatinsky condition with parameter in Λ, see [11].

The general answer to Question 1 for elliptic operators on manifolds with conical singularities has been open since Cheeger's seminal paper [2]. Cheeger's paper initiated considerable research in this direction with partial answers to Question 1 for a variety of particular cases, see for example [1,13–15,18]. The interest to answer the question for general elliptic cone operators has been renewed by the discovery of unusual or exotic behavior of the ζ-function for particular examples, see [3–5, 12, 13, 16]. Phrased in terms of the resolvent trace, what makes these examples unusual or exotic is that the asymptotic structure of the resolvent trace differs substantially from (3). Building on our previous work [6–9], we were able to answer Question 1 in [10] for general closed extensions of elliptic cone operators, see Theorem 1 below.

2 The Resolvent Trace of Elliptic Cone Operators

Let $\overline{M}$ be a compact manifold with boundary of dimension n, and let x be a defining function for $\partial\overline{M}$. The interior M of $\overline{M}$ is equipped with a c-metric ^cg, i.e., a Riemannian metric that in coordinates near the boundary is given by a smooth, positive definite cotensor in the forms dx and xdy_j up to $x = 0$. In other words, a c-metric is the Riemannian metric induced on M by any choice of metric on the c-cotangent bundle $^cT^*\overline{M}$. The latter is a vector bundle on $\overline{M}$ whose smooth sections are in one-to-one correspondence to all smooth 1-forms on $\overline{M}$ that are conormal to the boundary, see [6]. Locally near $Y = \partial\overline{M}$, a local frame for $^cT^*\overline{M}$ is given by the sections that correspond to the forms dx and xdy_j, where the y_j, $j = 1, \ldots, n-1$, are local coordinate maps on Y.

Let $E \to \overline{M}$ be a Hermitian vector bundle, and let $^cL^2(M; E)$ be the L^2-space with respect to ^cg and the Hermitian form on E. A cone operator is an operator $A \in x^{-m}\,\mathrm{Diff}_b^m(\overline{M}; E)$, where $\mathrm{Diff}_b^m(\overline{M}; E)$ is the space of totally characteristic differential operators of order m, see [17]. In coordinates near the boundary,

$$A = x^{-m} \sum_{k+|\alpha|\leq m} a_{k,\alpha}(x, y)(xD_x)^k D_y^\alpha \qquad (4)$$

with $a_{k,\alpha}(x, y)$ smooth up to $x = 0$. Here and in what follows we assume that $m > 0$. The following principal symbols are associated with A:

- The c-principal symbol $^c\varpi(A)$, defined on $^cT^*\overline{M} \setminus 0$. If A is as in (4) near the boundary, then
$$^c\varpi(A) = \sum_{k+|\alpha|=m} a_{k,\alpha}(x, y)\xi^k\eta^\alpha.$$

- The normal or model operator $A_\wedge : C_c^\infty(Y^\wedge; E_\wedge) \to C^\infty(Y^\wedge; E_\wedge)$, where $Y^\wedge$ is the inward pointing half of the normal bundle of $Y = \partial\overline{M}$, and $E_\wedge$ is the pull-back of $E|_{\partial\overline{M}}$ to $Y^\wedge$ with respect to the canonical projection. In coordinates we have
$$A_\wedge = x^{-m} \sum_{k+|\alpha|\leq m} a_{k,\alpha}(0, y)(xD_x)^k D_y^\alpha$$

if A is as in (4).

The function $x_\wedge = dx : NY \to \mathbb{R}$ trivializes $Y^\wedge$ as $Y \times \overline{\mathbb{R}}_+$. On $Y^\wedge$ we consider the Riemannian metric $dx_\wedge^2 + g_Y$ for any choice of metric g_Y on Y, and $E_\wedge$

carries the Hermitian metric induced by the metric on $E|_{\partial \overline{M}}$ and pull-back. Let $^cL^2(Y^\wedge; E_\wedge)$ be the associated L^2-space, and $\mathcal{D}_{\wedge,\min}$ and $\mathcal{D}_{\wedge,\max}$ be the domains of the minimal and maximal extension of $A_\wedge$ in that space, respectively. Under the assumption of c-ellipticity, i.e., the c-principal symbol $^c\varpi(A)$ is invertible on $^cT^*\overline{M} \setminus 0$, we constructed in [6] a canonical isomorphism

$$\theta : \mathcal{D}_{\max}(A)/\mathcal{D}_{\min}(A) \to \mathcal{D}_{\wedge,\max}/\mathcal{D}_{\wedge,\min}$$

that allows passage from the domains of closed extensions of A in $^cL^2(M; E)$ to those of $A_\wedge$ in $^cL^2(Y^\wedge; E_\wedge)$.

Theorem 1 ([10]). *Let $^c\varpi(A) - \lambda$ be invertible for λ in a closed sector $\Lambda \subsetneq \mathbb{C}$ which is a sector of minimal growth for $A_\wedge$ with the associated domain $\mathcal{D}_\wedge$ defined via $\mathcal{D}_\wedge/\mathcal{D}_{\wedge,\min} = \theta(\mathcal{D}/\mathcal{D}_{\min})$.*

Then Λ is a sector of minimal growth for $A_\mathcal{D}$, and for $\ell \in \mathbb{N}$ with $\ell > n/m$,

$$\mathrm{Tr}(A_\mathcal{D} - \lambda)^{-\ell} \sim \sum_{j=0}^{n-1} \alpha_j \lambda^{\frac{n-\ell m-j}{m}} + \alpha_n \log(\lambda)\lambda^{-\ell} + s_\mathcal{D}(\lambda) \text{ as } |\lambda| \to \infty$$

with coefficients $\alpha_j \in \mathbb{C}$ that are independent of the choice of domain $\mathcal{D}$, and a remainder $s_\mathcal{D}(\lambda)$ of order $O(|\lambda|^{-\ell})$. More precisely, in general we have an expansion

$$s_\mathcal{D}(\lambda) \sim \sum_{j=0}^{\infty} r_j(\lambda^{i\mu_1}, \ldots, \lambda^{i\mu_N}, \log \lambda)\lambda^{\nu_j/m} \text{ as } |\lambda| \to \infty, \tag{5}$$

where each r_j is a rational function in $N + 1$ variables, $N \in \mathbb{N}_0$, with real numbers μ_k, $k = 1, \ldots, N$, and $-\ell m \geq \nu_j > \nu_{j+1} \to -\infty$ as $j \to \infty$. We have $r_j = p_j/q_j$ with p_j, $q_j \in \mathbb{C}[z_1, \ldots, z_{N+1}]$ such that $q_j(\lambda^{i\mu_1}, \ldots, \lambda^{i\mu_N}, \log \lambda)$ is uniformly bounded away from zero for large λ.

The numbers μ_j and ν_j in (5) are determined by the boundary spectrum of A, more precisely the part of the boundary spectrum that directly relates to the $^cL^2$-extensions of A, see [10]. What terms actually occur in the asymptotic expansion depends strongly on certain dynamical properties of $\mathcal{D}_\wedge/\mathcal{D}_{\wedge,\min}$ with respect to the flow on the Grassmannian of subspaces of $\mathcal{D}_{\wedge,\max}/\mathcal{D}_{\wedge,\min}$ that is induced by the scaling action κ_ϱ on $Y^\wedge = Y \times \overline{\mathbb{R}}_+$; for functions $u(y, x)$ we have $\kappa_\varrho u(y, x) = u(y, \varrho x)$ for $\varrho > 0$ and $(y, x) \in Y \times \mathbb{R}_+$. This dynamical viewpoint is crucial for the proof and the understanding of Theorem 1.

Acknowledgements Work partially supported by the National Science Foundation, Grants DMS-0901173 and DMS-0901202.

References

1. J. Brüning and R. Seeley, *The resolvent expansion for second order regular singular operators*, J. Funct. Anal. **73** (2) (1987) 369–429.
2. J. Cheeger, *On the spectral geometry of spaces with cone-like singularities*, Proc. Nat. Acad. Sci. USA **76** (1979), 2103–2106.
3. H. Falomir, M.A. Muschietti, P.A.G. Pisani, *On the resolvent and spectral functions of a second order differential operator with a regular singularity*, J. Math. Phys. **45** (12) (2004) 4560–4577.
4. H. Falomir, M.A. Muschietti, P.A.G. Pisani, and R. Seeley, *Unusual poles of the ζ-functions for some regular singular differential operators*, J. Phys. A **36** (2003), no. 39, 9991–10010.
5. H. Falomir, P.A.G. Pisani, and A. Wipf, *Pole structure of the Hamiltonian ζ-function for a singular potential*, J. Phys. A **35** (2002), 5427–5444.
6. J. Gil, T. Krainer, and G. Mendoza, *Geometry and spectra of closed extensions of elliptic cone operators*, Canad. J. Math. **59** (2007), no. 4, 742–794.
7. _____, *Resolvents of elliptic cone operators*, J. Funct. Anal. **241** (2006), no. 1, 1–55.
8. _____, *On rays of minimal growth for elliptic cone operators*, Oper. Theory Adv. Appl. **172** (2007), 33–50.
9. _____, *Trace expansions for elliptic cone operators with stationary domains*, Trans. Amer. Math. Soc. **362** (2010), 6495–6522.
10. _____, *Dynamics on Grassmannians and resolvents of cone operators*, Anal. PDE **4** (2011), 115–148.
11. G. Grubb, *Functional calculus of pseudodifferential boundary problems*, 2nd ed., Progress in Mathematics, vol. 65. Birkhäuser, Basel, 1996.
12. K. Kirsten, P. Loya, and J. Park, *The very unusual properties of the resolvent, heat kernel, and zeta function for the operator $-d^2/dr^2 - 1/(4r^2)$*, J. Math. Phys. **47** (2006), no. 4, 043506, 27 pp.
13. _____, *Exotic expansions and pathological properties of ζ-functions on conic manifolds*, J. Geom. Anal. **18** (2008), no. 3, 835–888.
14. M. Lesch, *Operators of Fuchs type, conical singularities, and asymptotic methods*, Teubner-Texte zur Math. vol 136, B.G. Teubner, Stuttgart, Leipzig, 1997.
15. P. Loya, *Parameter-dependent operators and resolvent expansions on conic manifolds*, Illinois J. Math. **46** (2002), no. 4, 1035–1059.
16. P. Loya, P. McDonald, and J. Park, *Zeta regularized determinants for conic manifolds*, J. Funct. Anal. **242** (2007), no. 1, 195–229.
17. R. Melrose, *The Atiyah-Patodi-Singer index theorem*, Research Notes in Mathematics, A K Peters, Ltd., Wellesley, MA, 1993.
18. E. Mooers, *Heat kernel asymptotics on manifolds with conic singularities*, J. Anal. Math. **78** (1999) 1–36.
19. R. Seeley, *Complex powers of an elliptic operator*, Singular Integrals, AMS Proc. Symp. Pure Math. X, 1966, Amer. Math. Soc., Providence, 1967, pp. 288–307.

Spectral Geometry for the Riemann Moduli Space

Rafe Mazzeo

A report on joint work with Lizhen Ji, Werner Müller and Andras Vasy.

It is only within the last few decades that the study of analytic and geometric problems in the setting of spaces with structured singularities has become a real focus in geometric analysis and partial differential equations. Unlike for the many traditional problems concerning spaces and equations with overall low regularity, the focus here is on spaces which are mostly smooth, but which have stratified singular sets and metrics (or coefficients of the PDE) which are adapted to these singularities. Because of this, one can expect much sharper results. There are many reasons for studying such spaces, chief amongst which is that from many points of view they are just as natural as smooth manifolds. Examples of this type of singular space appear in the study of algebraic and analytic varieties, configuration and moduli spaces, and even just as level sets of smooth functions (Morse-Bott theory). Granting the interest of these spaces, one then wishes to extend the techniques, methods and goals of geometric analysis to these settings.

There are several important themes in this subject. One revolves around the development of systematic techniques to study, for example, elliptic differential operators on spaces with various types of singularities. Even the simplest type of singularity, namely compact manifolds with isolated conic singularities, has provided a very rich field of study, with many interesting new developments. Generalizations include the study of manifolds with simple edge singularities, or more generally still, iterated edge singularities, sometimes also known as smoothly stratified spaces. There is now a body of work even in this relatively general setting, beginning with the foundational work of Cheeger [2], but see also [1, 3, 5, 6].

A related theme concerns the study of very specific singular spaces which arise "in nature", preferably with certain canonical metrics on them. Here one should include the enormous body of work concerning analysis on singular projective

R. Mazzeo (✉)
Department of Mathematics, Stanford University, Stanford, CA, 94305, USA
e-mail: mazzeo@math.stanford.edu

D. Grieser et al. (eds.), *Microlocal Methods in Mathematical Physics and Global Analysis*, Trends in Mathematics, DOI 10.1007/978-3-0348-0466-0_15, © Springer Basel 2013

varieties, on compactifications of locally symmetric spaces, and on singular moduli spaces. One of the most classical objects of this type is the Riemann moduli space $\mathcal{M}_g$ of conformal structures (or equivalently, hyperbolic metrics) on a fixed compact Riemann surface of genus $g > 1$, moduli diffeomorphisms. This space is commonly represented as the quotient $\mathcal{T}_g/\mathrm{Map}(\Sigma)$, where the mapping class group, $\mathrm{Map}(\Sigma) = \mathrm{Diff}(\Sigma)/\mathrm{Diff}_0(\Sigma)$, is the quotient of the space of all diffeomorphisms of Σ by the closed subgroup of all diffeomorphisms which are isotopic to the identity; the space $\mathcal{T}_g$ is the Teichmüller space, which is the quotient of the space of all hyperbolic metrics on Σ by $\mathrm{Diff}_0(\Sigma)$.

The Deligne-Mumford compactification $\overline{\mathcal{M}}_g$ of this space has a somewhat complicated singular structure, but with certain simplifying features. It is a complex space, singular along the union of a collection $D_0, \ldots, D_{[g/2]}$ of immersed divisors with simple normal crossings, and carries several important metrics. We choose to focus on two of these: the Weil-Petersson metric, which is the natural L^2 metric on the tangent space, and the so-called Ricci metric (or rather, certain mollifications of it). The former is incomplete, and an instance of an 'iterated cusp-edge' metric, while the latter is complete and has a fairly simple asymptotic product structure at infinity. A good survey of current knowledge about the former of these metrics is contained in [7], and we refer there for a much more extensive bibliography.

The work reported in this talk is joint with Lizhen Ji, Werner Müller and Andras Vasy. Our goal is analyze the Laplacian and other natural elliptic operators for the Weil-Petersson metric from the point of view of spectral geometry, and similarly to analyze the corresponding operators for the smoothed Ricci metric from the point of view of geometric scattering theory. The former of these projects is now at a more advanced stage, so we restrict discussion here to that case.

To set the stage, we recall what is known about the asymptotics of the Weil-Petersson metric g_{WP} near the singular divisors. The approximate structure of this metric was first obtained by Masur in the 1970s [4]; this was later substantially sharpened by Yamada [8], with further results on its structure by Wolpert [7]. The upshot of these papers is that if p is a point in the singular set, then there is a local set of holomorphic coordinates $(z_1, \ldots, z_n$, $n = 3g - 3$, in some punctured neighbourhood, such that if we write $z_j = r_j e^{i\theta_j}$, then

$$g_{\mathrm{WP}} = \sum_{j=1}^{k} (dr_j^2 + r^6 d\theta_j^2)(1 + \mathcal{O}(r^3)) + g_D + k;$$

here $r = |(r_1, \ldots, r_k)|$ and g_D is some induced metric on the singular stratum where $z_1 = \ldots = z_k = 0$; the final term k is a higher order remainder term which is irrelevant for our purposes below. Before carrying out more refined spectral geometric analysis of this space, it will be necessary to establish higher order asymptotics of g_{WP}, but these are not yet known.

We announce some basic results about the scalar Laplacian Δ for the Weil-Petersson metric.

Theorem 1. *The operator Δ acting on $\mathcal{C}_0^\infty(\mathcal{M}_g)$ has a unique self-adjoint extension to an unbounded operator on $L^2(\overline{\mathcal{M}_g})$, which we continue to denote by Δ. Furthermore, this operator has discrete spectrum $\{\lambda_j\}$, and the spectral counting function $N(\lambda) = \#\{j : \lambda_j \leq \lambda\}$ satisfies the Weyl law*

$$N(\lambda) = \frac{\omega_{6g-6}}{(2\pi)^{6g-6}} \mathrm{Vol}(\mathcal{M}_g) \lambda^{n/2} + o(\lambda^{n/2}).$$

(where ω_ℓ is the volume of the Euclidean ball B^ℓ).

This sets the stage for further investigations: one expects interesting connections between $\mathrm{spec}(\Delta)$ and other aspects of the geometry of this space.

Essential self-adjointness is obtained by showing that there are no L^2 solutions to $(\Delta \pm i)u = 0$. These can be ruled out if we prove the standard identity $\langle -\Delta u, u \rangle = \|\nabla u\|^2$ without additional boundary terms. In other words, it suffices to prove that if u and Δu both lie in L^2, then we can control the growth of u near the singular set enough to carry out this integration by parts. This is done using an elaboration of the Hardy inequality. The discreteness of the spectrum is then proved by showing that the (now unique) domain $\mathrm{Dom}(\Delta)$ is compactly contained in L^2. Finally, to obtain the Weyl law in this relatively crude form (i.e. with no estimate of the remainder), we can use standard comparison techniques (Dirichlet-Neumann bracketing), once we have verified that an arbitrarily small neighbourhood of the singular set contributes a lower order term. Amongst these three arguments, the first one is more difficult than the others.

There are many further directions. Current work of Gell-Redman proves a full asymptotic expansion for the heat kernel associated to metrics with the same 'crossing cubic cuspidal' structure as g_{WP}, but assuming that these metrics themselves have full asymptotic regularity. The asymptotics of the heat trace involve some new and potentially interesting terms. One significant goal is to obtain a signature formula for $(\overline{\mathcal{M}_g}, g_{WP})$, and this heat kernel analysis should provide a crucial tool for this. However, it remains to show that g_{WP} itself does indeed have full asymptotic regularity or at least that what is known about its asymptotic structure suffices to understand enough of the expansion of the heat trace to obtain such index formulas.

References

1. Albin, P., Leichtnam, E., Mazzeo, R., Piazza, P. "The signature package on Witt spaces". To appear, Ann. Sci. de l'Ecole Norm. Sup.
2. Cheeger, J. "Spectral geometry of singular Riemannian spaces" Jour. Diff. Geom. 18 (1983) No. 4, 575–657.
3. Gil, J., Krainer, T., Mendoza, G. "On the closure of elliptic wedge operators" arXiv:1007.2397
4. Masur, H. "Extension of the Weil-Petersson metric to the boundary of Teichmüller space" Duke Math. J. 43 (3) (1976), 623–635.

5. Mazzeo, R. "Elliptic theory of differential edge operators, I" Comm. Par. Diff. Eqns. 16 (1991)
   No. 10, 1615–1664.
6. Melrose, R.B. "The Atiyah-Patodi-Singer index theorem" Research Notes in Mathematics, 4.
   A.K. Peters, Ltd. Wellesley, MA (1993).
7. Wolpert, S. "The Weil-Petersson metric geometry" in Handbook of Teichmüller theory Vol. II,
   47–64. IRMA Lect. Math. Theor. Phys. 13, Eur. Math. Soc., Zürich (2009).
8. Yamada, S. "On the geometry of Weil-Petersson completion of Teichmüller spaces" Math. Res.
   Lett. 10 (2–3) (2003), 391–400.

Invariant Integral Operators on the Oshima Compactification of a Riemannian Symmetric Space: Kernel Asymptotics and Regularized Traces

Pablo Ramacher and Aprameyan Parthasarathy

Let $\mathbb{X}$ be a Riemannian symmetric space of non-compact type. Then $\mathbb{X}$ is isomorphic to G/K, where G is a connected, real, semisimple Lie group, and K a maximal compact subgroup. Consider further the Oshima compactification $\widetilde{\mathbb{X}}$ of $\mathbb{X}$ [8], which is a simply connected, closed, real-analytic manifold carrying an analytic G-action. The orbital decomposition of $\widetilde{\mathbb{X}}$ is of normal crossing type, and the open orbits are isomorphic to G/K, the number of them being equal to 2^l, where l denotes the rank of G/K. We study invariant integral operators of the form

$$\pi(f) = \int_G f(g)\pi(g)d_G(g), \tag{1}$$

where π is the regular representation of G on the Banach space $\mathrm{C}(\widetilde{\mathbb{X}})$ of continuous functions on $\widetilde{\mathbb{X}}$, f a smooth, rapidly decreasing function on G, and d_G a Haar measure on G. These operators play an important role in representation theory, and our interest will be directed towards the elucidation of the microlocal structure of the operators $\pi(f)$. Since the underlying group action on $\widetilde{\mathbb{X}}$ is not transitive, the operators $\pi(f)$ are not smooth, and the orbit structure of $\widetilde{\mathbb{X}}$ is reflected in the singular behavior of their Schwartz kernels. As it turns out, the operators in question can be characterized as totally characteristic pseudodifferential operators, a class which was first introduced in [7] in connection with boundary problems. In fact, if $\widetilde{\mathbb{X}}_\Delta$ denotes a component in $\widetilde{\mathbb{X}}$ isomorphic to G/K, we prove that the restrictions

$$\pi(f)_{|\overline{\widetilde{\mathbb{X}}_\Delta}} : \mathrm{C}_c^\infty(\overline{\widetilde{\mathbb{X}}_\Delta}) \longrightarrow \mathrm{C}^\infty(\overline{\widetilde{\mathbb{X}}_\Delta})$$

P. Ramacher (✉) · A. Parthasarathy
Fachbereich Mathematik und Informatik, Philipps-Universität Marburg, Hans-Meerwein-Strasse, 35032, Marburg, Germany
e-mail: ramacher@mathematik.uni-marburg.de; apra@mathematik.uni-marburg.de

D. Grieser et al. (eds.), *Microlocal Methods in Mathematical Physics and Global Analysis*, 73
Trends in Mathematics, DOI 10.1007/978-3-0348-0466-0_16, © Springer Basel 2013

of the operators $\pi(f)$ to the manifold with corners $\overline{\widetilde{\mathbb{X}_\Delta}}$ are totally characteristic pseudodifferential operators of class $L_b^{-\infty}$. A similar structure theorem for invariant integral operators on prehomogeneous vector spaces was already obtained in [9].

As a special class of integral operators, we consider the holomorphic semigroup generated by a strongly elliptic operator Ω associated to the regular representation $(\pi, \mathrm{C}(\widetilde{\mathbb{X}}))$ of G, as well as its resolvent. Since both the holomorphic semigroup and the resolvent can be characterized as operators of the form (1), they can be studied with the previous methods, and relying on the theory of elliptic operators on Lie groups [10] we obtain a description of the asymptotic behavior of the semigroup and resolvent kernels on $\widetilde{\mathbb{X}}_\Delta \simeq \mathbb{X}$ at infinity. In the particular case of the Laplace-Beltrami operator on $\mathbb{X}$, these questions have been studied before. For the classical heat kernel on $\mathbb{X}$, precise upper and lower bounds were previously obtained in [1] using spherical analysis, while a detailed description of the analytic properties of the resolvent of the Laplace-Beltrami operator on $\mathbb{X}$ was given in [5] and [6].

Using the structure theorem, a regularized trace for the operators $\pi(f)$ can be defined, yielding a distribution on the group G which is defined to be the character of the representation $(\pi, C(\widetilde{\mathbb{X}}))$. In fact, in his early work on infinite dimensional representations of semi-simple Lie groups, Harish–Chandra [4] realized that the correct generalization of the character of a finite-dimensional representation was a distribution on the group given by the trace of a convolution operator on representation space. This distribution character is given by a locally integrable function which is analytic on the set of regular elements, and satisfies character formulae analogous to the finite dimensional case. Later, Atiyah and Bott [2] gave a similar description of the character of a parabolically induced representation in their work on Lefschetz fixed point formulae for elliptic complexes. More precisely, let H be a closed, co-compact subgroup of G, and ϱ a representation of H on a finite dimensional vector space V. If $T(g) = (\iota_*\varrho)(g)$ is the representation of G induced by ϱ in the space of sections over G/H with values in the homogeneous vector bundle $G \times_H V$, then its distribution character is given by the distribution

$$\Theta_T : \mathrm{C}_c^\infty(G) \ni f \longmapsto \mathrm{Tr}\, T(f), \qquad T(f) = \int_G f(g)T(g)d_G(g),$$

where d_G denotes a Haar measure on G. The point to be noted is that $T(f)$ is a smooth operator, and since G/H is compact, it does have a well-defined trace. On the other hand, assume that $g \in G$ is transversal, meaning that it acts on G/H only with simple fixed points. In this case, a transversal trace $\mathrm{Tr}^\flat T(g)$ of $T(g)$ can be defined within the framework of pseudodifferential operators, which is given by a sum over fixed points of g. Atiyah and Bott then showed that, on an open set $G_T \subset G$ of transversal elements,

$$\Theta_T(f) = \int_{G_T} f(g)\,\mathrm{Tr}^\flat T(g)d_G(g), \qquad f \in \mathrm{C}_c^\infty(G_T).$$

This means that, on G_T, the character Θ_T of the induced representation T is represented by the locally integrable function $\mathrm{Tr}^\flat \, T(g)$, and its computation reduced to the evaluation of a sum over fixed points. When G is a p-adic reductive group defined over a non-Archimedean local field of characteristic zero, a similar analysis of the character of a parabolically induced representation was carried out in [3]. In our case, the convolution operators $\pi(f)$, where $f \in C_c^\infty(G)$, are not smooth, and therefore do not have a well-defined trace. Nevertheless, using the fact that they can be characterized as totally characteristic pseudodifferential operators of order $-\infty$, we are able to define a regularized trace $\mathrm{Tr}_{reg} \, \pi(f)$ for the operators $\pi(f)$, and in this way obtain a map

$$\Theta_\pi : C_c^\infty(G) \ni f \mapsto \mathrm{Tr}_{reg}(f) \in \mathbb{C},$$

which is shown to be a distribution on G. This distribution is defined to be the character of the representation π. We then show that, on a certain open set $G(\widetilde{\mathbb{X}})$ of transversal elements,

$$\mathrm{Tr}_{reg} \, \pi(f) = \int_{G(\widetilde{\mathbb{X}})} f(g) \, \mathrm{Tr}^\flat \, \pi(g) d_G(g), \qquad f \in C_c^\infty(G(\widetilde{\mathbb{X}})),$$

where, with the notation $\Phi_g(\tilde{x}) = g \cdot \tilde{x}$,

$$\mathrm{Tr}^\flat \, \pi(g) = \sum_{\tilde{x} \in \mathrm{Fix}(\widetilde{\mathbb{X}}, g)} \frac{1}{|\det(\mathbf{1} - d\,\Phi_g(\tilde{x}))|},$$

the sum being over the (simple) fixed points of $g \in G(\widetilde{\mathbb{X}})$ on $\widetilde{\mathbb{X}}$. Thus, on the open set $G(\widetilde{\mathbb{X}})$, Θ_π is represented by the locally integrable function $\mathrm{Tr}^\flat \, \pi(g)$, which is given by a formula similar to the character of a parabolically induced representation. It is likely that similar distribution characters could be introduced for G-manifolds with a dense union of open orbits, or for spherical varieties, and that corresponding character formulae could be proved.

References

1. J.-P. Anker and L. Ji, *Heat kernel and Green function estimates on noncompact symmetric spaces*, Geom. Funct. Anal. **9** (1999), no. 6, 1035–1091.
2. M.F. Atiyah and R. Bott, *A Lefschetz fixed point formula for elliptic complexes: II. Applications*, Ann. of Math. **88** (1968), 451–491.
3. L. Clozel, *Théoréme d'Atiyah-Bott pour les variétés p-adiques et caractéres des groupes réductifs*, Mem. de la S.M.F. **15** (1984), 39–64.
4. Harish-Chandra, *Representations of semisimple Lie groups. III*, Trans. Am. Math. Soc. **76** (1954), 234–253.
5. R. R. Mazzeo and R. B. Melrose, *Meromorphic extension of the resolvent on complete spaces with asymptotically constant negative curvature*, J. Funct. Anal. **75** (1987), 260–310.

6. R. R. Mazzeo and A. Vasy, *Analytic continuation of the resolvent of the Laplacian on symmetric spaces of noncompact type*, J. Funct. Anal. **228** (2005), 311–368.
7. R. Melrose, *Transformation of boundary problems*, Acta Math. **147** (1982), 149–236.
8. T. Oshima, *A realization of Riemannian symmetric spaces*, J. Math. Soc. Japan **30** (1978), no. 1, 117–132.
9. P. Ramacher, *Pseudodifferential operators on prehomogeneous vector spaces*, Comm. Partial Diff. Eqs. **31** (2006), 515–546.
10. D. W. Robinson, *Elliptic operators and Lie groups*, Oxford University Press, Oxford, 1991.

Pseudodifferential Operators on Manifolds with Foliated Boundaries

Frédéric Rochon

1991 Mathematics Subject Classification: 58J28, 58J32, 58J40.

1 Manifolds with Fibred Boundaries

Let X be a compact manifold with boundary ∂X endowed with a fibration

$$Z \longrightarrow \partial X$$
$$\downarrow \Phi$$
$$Y.$$

Let $x \in \mathcal{C}^\infty(X)$ be a boundary defining function and let g_Φ be a complete Riemannian metric on $X \setminus \partial X$ which in a collar neighborhood of ∂X is of the form

$$g_\Phi = \frac{dx^2}{x^4} + \frac{\Phi^* h}{x^2} + \kappa, \tag{1}$$

where κ is a symmetric 2-tensor restricting to give a metric on each fibre of Φ and h is a Riemannian metric on Y. To study geometric operators (Laplacian, Dirac operators) associated to such metrics, Mazzeo and Melrose introduced a calculus of

F. Rochon (✉)
Université de Québec á Montréal, Montreal, Canada
e-mail: rochon.frederic@uqam.ca

pseudodifferential operators: the Φ-calculus. The starting point is the Lie algebra of Φ-vector fields:

$$\mathcal{V}_{\Phi}(X) = \{\xi \in \Gamma(TX) \mid \xi x \in x^2 C^{\infty}(X), \ \Phi_*(\xi|_{\partial X}) = 0\}.$$

In local coordinates, such a vector $\xi \in \mathcal{V}_{\Phi}(X)$ takes the form:

$$\xi = ax^2 \frac{\partial}{\partial x} + \sum_i b^i x \frac{\partial}{\partial y^i} + \sum_j c^j \frac{\partial}{\partial z^j}, \quad a, b^i, c^j \in C^{\infty}(X).$$

Since it is a Lie algebra, we can consider its universal enveloping algebra to define Φ-differential operators. Mazzeo and Melrose defined more generally Φ-pseudodifferential operators. They are useful to study mapping properties, for instance to determine when a Φ-differential operator is Fredholm.

2   Manifolds with Foliated Boundaries

Question 1. What can we do when the fibration Φ is replaced by a smooth foliation $\mathcal{F}$ on ∂X?

The notion of $\mathcal{F}$-vector fields is easy to define:

$$\mathcal{V}_{\mathcal{F}}(X) = \{\xi \in \Gamma(TX) \mid \xi x \in x^2 C^{\infty}(X), \ \xi|_{\partial X} \in \Gamma(T\mathcal{F})\}.$$

This is still a Lie algebra, so we can define $\mathcal{F}$-differential operators. However, since pseudodifferential operators are not local, we expect global aspects of the foliation $\mathcal{F}$ to come into play. One approach consists in using groupoid theory, namely, since $\mathcal{V}_{\mathcal{F}}(X)$ is in fact a Lie algebroid, we can integrate it to get a Lie groupoid $\mathcal{G}$. We can then use the general approach of Nistor-Weinstein-Xu to construct a pseudodifferential calculus. We will instead proceed differently by assuming the foliation can be 'resolved' by a fibration. This restricts the class of foliations that can be considered, but will allow us to develop further the underlying analysis.

We will assume the foliation arises as follows:

1. $\partial X = \partial \widetilde{X}/\Gamma$, where Γ is a discrete group acting freely and properly discontinuously on $\partial \widetilde{X}$, a possibly non-compact manifold;
2. There is a fibration $\Phi : \partial \widetilde{X} \to Y$ with Y a compact manifold;
3. The group Γ acts Y in a locally free manner (that is, if $\gamma \in \Gamma$ and $\mathcal{U} \subset Y$ an open set are such that $y \cdot \gamma = y$ for all $y \in \mathcal{U}$, then γ is the identity element) and so that $\Phi(p \cdot \gamma) = \Phi(p) \cdot \gamma$ for all $p \in \partial \widetilde{X}$ and $\gamma \in \Gamma$;
4. The images of the fibres of Φ under the quotient map $q : \partial \widetilde{X} \to \partial X$ give the leaves of the foliation $\mathcal{F}$.

Example 1. The Kronecker foliation on the 2-torus with lines of irrational slope θ arise in this way. One takes $\partial \widetilde{X} = \mathbb{R} \times \mathbb{R}/\mathbb{Z}$ with the fibration Φ given by the projection on the right factor $Y = \mathbb{R}/\mathbb{Z}$, and the group Γ to be the integers with action given by

$$(x, [y]) \cdot k = (x + k, [y - k\theta]), \quad [y] \cdot k = [y - k\theta], \quad k \in \mathbb{Z}.$$

The identification with the standard definition of the Kronecker foliation is then given by the map

$$\Psi : (\mathbb{R} \times \mathbb{R}/\mathbb{Z})/\mathbb{Z} \to \mathbb{T}^2 = \mathbb{R}/\mathbb{Z} \times \mathbb{R}/\mathbb{Z}$$
$$[x, [y]] \quad \mapsto \quad ([x], [y + \theta x]).$$

Example 2. Seifert fibrations (circle foliations on a compact 3-manifold) typically arise in this way, except when the space of leaves is a bad orbifold.

For such foliations, we can define $\mathcal{F}$-operators as follows. We let $M = \partial X \times [0, \epsilon)_x \subset X$ be a collar neighborhood of ∂X and consider $\widetilde{M} = \partial \widetilde{X} \times [0, \epsilon)_x$ with Γ acting on $\widetilde{M}$ in obvious way so that $\widetilde{M}/\Gamma = M$. On $\widetilde{M}$, we consider the space of Γ-invariant Φ-operators $\Psi^*_{\Phi,\Gamma}(\widetilde{M})$ with support away from $x = \epsilon$. Given $\widetilde{P} \in \Psi^k_{\Phi,\Gamma}(\widetilde{M})$, we can make it act on $f \in \mathcal{C}^\infty(M)$ by requiring that $\widetilde{P}(q^* f) = q^* \widetilde{P}(f)$, where $q : \widetilde{M} \to M$ is the quotient map. This is meaningful because $\widetilde{P}$ acts on Γ invariant functions to give again Γ-invariant functions. We denote by $q_* \widetilde{P}$ the operator acting on $\mathcal{C}^\infty(M)$ obtained from $\widetilde{P}$ in this way.

Definition 1. An $\mathcal{F}$-pseudodifferential operator $P \in \Psi^m_{\mathcal{F}}(X)$ is an operator of the form

$$P = q_* P_1 + P_2, \quad P_1 \in \Psi^m_{\Phi,\Gamma}(\widetilde{M}), \ P_2 \in \dot{\Psi}^m(X).$$

From the Φ-calculus, we deduce relatively easily that $\mathcal{F}$-operators are closed under composition, that they map smooth functions to smooth functions and that they are bounded when acting on appropriate Sobolev spaces. One can also introduce a notion of principal symbol $\sigma_m(P)$ as well as a notion of normal operator $N_{\mathcal{F}}(P)$ defined by 'restricting' the operator P to the boundary. This leads to a simple criterion to describe Fredholm operators. An operator $P \in \Psi^m_{\mathcal{F}}(X)$ is Fredholm (when acting on suitable Sobolev spaces) if and only if its principal symbol $\sigma_m(P)$ and its normal operator $N_{\mathcal{F}}(P)$ are invertible.

3 An Index Theorem for Some Dirac-Type Operators

Assume now that the the foliation $\mathcal{F}$ is also such that $\partial \widetilde{X}$ is compact and the group Γ is finite. In particular, the leaves of $\mathcal{F}$ must be compact. Let $g_{\mathcal{F}}$ be a metric such that $q^*(g_{\mathcal{F}}|_M)$ takes the form (1) near $\partial \widetilde{X}$. Suppose X is even dimensional and that $X, \widetilde{X}$ and Y are spin manifolds. Let $D_{\mathcal{F}}$ be the induced Dirac operator. Suppose its normal operator $N_{\mathcal{F}}(D_{\mathcal{F}})$ is invertible, which is the case for instance when the induced metric on the leaves of the foliation $\mathcal{F}$ has positive scalar curvature. Under the decomposition $S = S^+ \oplus S^-$ of the spinor bundle, the Dirac operator can be written as

$$D_{\mathcal{F}} = \begin{pmatrix} 0 & D_{\mathcal{F}}^- \\ D_{\mathcal{F}}^+ & 0 \end{pmatrix}.$$

Since $N_{\mathcal{F}}(D_{\mathcal{F}})$ is invertible, the operator $D_{\mathcal{F}}^+$ is Fredholm.

Theorem 1. *The index of $D_{\mathcal{F}}^+$ is given by*

$$\mathrm{ind}(D_{\mathcal{F}}^+) = \int_X \widehat{A}(X, g_{\mathcal{F}}) - \frac{1}{|\Gamma|} \int_Y \widehat{A}(Y, h) \widehat{\eta}(\widetilde{D}_0) + \frac{\rho}{2},$$

where $\widetilde{D}_0$ is a family of Dirac operators on the fibres of $\Phi : \partial\widetilde{X} \to Y$ associated to $q^*(D_{\mathcal{F}}|_M)$ and $\rho = \frac{\eta(\widetilde{D}_\delta)}{|\Gamma|} - \eta(D_\delta)$ is a difference of two eta invariants with $\widetilde{D}_\delta$ the Dirac operator on $(\partial\widetilde{X}, \frac{\Phi^*h}{\delta^2} + \kappa)$ and D_δ the Dirac operator on $(\partial X, q_*(\frac{\Phi^*h}{\delta^2} + \kappa))$. Both $\widetilde{D}_\delta$ and D_δ are invertible for $\delta > 0$ small enough and ρ does not depend on δ.

The strategy to prove this theorem is to take an adiabatic limit.

The Determinant of the Laplacian on a Conically Degenerating Family of Metrics

David A. Sher

One way to approach analysis problems on singular manifolds is to approximate by a degenerating sequence of smooth manifolds and then to analyze the behavior of the relevant quantities under the degeneration. To this end, we consider a family of smooth manifolds Ω_ϵ degenerating to a manifold Ω_0 with a single conical singularity, and analyze the determinant of the Laplacian.

One motivation for this analysis is the work of Osgood, Phillips, and Sarnak in [5–7], who used the determinant to prove compactness of isospectral sets of closed surfaces and of planar domains. Khuri, a student of Sarnak, tried to extend the results to surfaces of arbitrary genus with arbitrary number of holes [3]. A key ingredient in [6] and [7] is the properness of the determinant on the relevant moduli space of constant-curvature surfaces. However, Khuri showed that this properness is untrue in the setting of flat tori with one hole, and the troublesome families of surfaces look quite a bit like a conical degeneration. We hope that our work will help us understand exactly what goes wrong here. Another motivation is that we would like to understand what happens to the analytic torsion, which is an alternating sum of determinants; a goal of Dai, Mazzeo, myself, and Vertman (among others) is to prove an analog of the Cheeger-Muller theorem on manifolds with conical singularities, and we hope that analyzing the analytic torsion on such a degenerating family yields insights into the relationship between the smooth and conic settings.

We recall the definition of the determinant, which can be found in [9] and elsewhere and is originally due to Ray and Singer [8]. Given a smooth compact manifold M with Laplacian Δ_M, we define

$$\zeta_M(s) = \frac{1}{\Gamma(s)} \int_0^\infty Tr(e^{-t\Delta_M} - P_0)t^{s-1}\,dt,$$

D.A. Sher (✉)
McGill University, Montreal, Canada
e-mail: dsher@stanford.edu

D. Grieser et al. (eds.), *Microlocal Methods in Mathematical Physics and Global Analysis*, Trends in Mathematics, DOI 10.1007/978-3-0348-0466-0_18, © Springer Basel 2013

where P_0 is the projector onto the constants. It is standard from the theory of short-time heat asymptotics (described in [9]) that $\zeta_M(s)$ has a meromorphic continuation to all of $\mathbb{C}$, with $s = 0$ as a regular value. Thus we can define $\zeta'_M(0)$, and we write

$$\det \Delta_M = e^{-\zeta'_M(0)}.$$

We also note that it is possible, with a few additional wrinkles, to define the determinant on smooth manifolds with isolated conical singularities [1]. More subtly, it is also possible to define a renormalized zeta function and determinant on an asymptotically conic manifold Z. The heat kernel is no longer trace class, but we instead replace $Tr(e^{-t\Delta_M})$ in the zeta function with a renormalized trace

$$^R Tr(e^{-t\Delta_Z}) = f.p._{\cdot\epsilon=0} \int_{r<1/\epsilon} H^Z(t,z,z)\, dz.$$

This is the finite part of a divergent expansion, analogous to Melrose's b-trace [4].

In our particular setting, we assume that Ω_0 has a point p and a neighborhood of p that is isometric to $[0,1]_r \times N^{n-1}$, with the conic metric $dr^2 + r^2 d\theta^2$, where (N, θ) is a closed Riemannian manifold of dimension $n - 1$ (the cross-section). To construct Ω_ϵ, we introduce a model space. Let Z be a complete manifold with a neighborhood of infinity isometric to $[1, \infty)_r \times N$, again with the conic metric. Note that in particular, Z is asymptotically conic. For each $\epsilon < 1$, we remove the tip of Ω_0 (cutting at $r = 1$) and replace it by a scaled copy of the tip of Z, namely $\epsilon(Z \cap \{r < 1/\epsilon\})$. One easily sees that this gluing process gives a smooth manifold Ω_ϵ. In fact, this is a special case of the "asymptotically conic convergence" defined by Rowlett in [10]. In this setting, we have proven the following:

Theorem 1. *As $\epsilon \to 0$, we have the divergent expansion*

$$\zeta'_{\Omega_\epsilon}(0) \sim^R \zeta_Z(0)(2\log\epsilon) +^R \zeta'_Z(0) + \zeta'_{\Omega_0}(0),$$

in the sense that the difference converges to zero as $\epsilon \to 0$.

In order to obtain this theorem, we consider the heat trace on Ω_ϵ, $TrH^{\Omega_\epsilon}(t)$. This is a function of ϵ and t and hence lives on the quadrant $Q = \{t \geq 0, \epsilon \geq 0\}$. Let Q_0 be the space obtained from Q by performing a radial blow-up in the coordinates $(\epsilon, \sqrt{t})$ at $\sqrt{t} = \epsilon = 0$, with blowdown map β. Then we claim that:

Theorem 2. $\beta^*(TrH^{\Omega_\epsilon}(t))$ *is polyhomogeneous conormal on Q_0.*

With this, as well as some information about the coefficients in the expansions, we can analyze the zeta function and determinant directly and prove Theorem 1.

The way we prove the structure theorem is via a direct parametrix construction of H^{Ω_ϵ}. Note that Ω_ϵ is equal to Ω_0 outside $r = 1$ and equal to ϵZ inside $r = 1$, so we define our initial guess G_ϵ by gluing together those heat kernels in the respective regions. We need to understand the structure of $G_\epsilon(t, z, z')$; there is a contribution from $H^{\Omega_0}(t, z, z')$ and also a contribution from $H^{\epsilon Z}(t, z, z')$. Using the scaling property that

$$H^{\epsilon Z}(t, z, z') = \epsilon^{-n} H^Z(t/\epsilon^2, z/\epsilon, z'/\epsilon),$$

we see that to understand the parametrix at time t, we need to understand the heat kernel on Z at the rescaled time t/ϵ^2. In particular, we need to understand the structure of the heat kernel on Z for large time as well as short time.

To do this, we exploit the relationship between the heat kernel and the resolvent, given by:

$$H^Z(t, z, z') = \int_\Gamma e^{-t\lambda} (\Delta + \lambda)^{-1}(z, z') \, d\lambda,$$

where λ is a contour around the spectrum. By making the change of variables $\lambda' = t\lambda$, we can see that the heat kernel at time t is analogous to the resolvent at energy $\lambda = 1/t$. So we need to understand the resolvent at low energy. In recent work, [2], Guillarmou and Hassell have analyzed this structure very thoroughly. By using and extending their results, we can describe the heat kernel on Z at long time as polyhomogeneous conormal on a particular blown-up space, which lets us understand the structure of G_ϵ. We then perform the usual parametrix construction, and by some careful analysis we can prove Theorem 2.

As a follow-up to this work, we would like to understand the analytic torsion, and to do this we need to get some version of Theorem 1 for the Laplacian on differential forms. If we follow this program, that means getting an analogue of Guillarmou and Hassell's results for the Laplacian on forms. However, in this case, there are additional difficulties because of the possible presence of harmonic forms.

Acknowledgements I would like to thank my Ph.D. advisor, Rafe Mazzeo, for introducing me to this problem and for all his support. I would also like to thank the organizers, D. Grieser, S. Teufel, and A. Vasy, for giving me the chance to speak at the conference.

References

1. J. CHEEGER, *Spectral geometry of singular Riemannian spaces*, J. Diff. Geom. **18** (1984) 575–657.
2. C. GUILLARMOU and A. HASSELL, *Resolvent at low energy and Riesz transform for Schrodinger operators on asymptotically conic manifolds, I*, Math. Ann. **341** (2008), 859896
3. H. KHURI, *Heights on the moduli space of Riemann surfaces with circle boundaries*, Duke Math. J. **64** (1991) 555–570.
4. R. MELROSE, *The Atiyah-Patodi-Singer Index Theorem*, Res. Notes Math. **4**, A. K. Peters, Wellesly, Mass., 1993.
5. B. OSGOOD, R. PHILLIPS and P. SARNAK, *Extremals of determinants of Laplacians*, J. Funct. Anal. **80** (1988), 148–211.
6. B. OSGOOD, R. PHILLIPS and P. SARNAK, *Compact isospectral sets of surfaces*, J. Funct. Anal **80** (1988), 212–234.
7. B. OSGOOD, R. PHILLIPS and P. SARNAK, *Moduli space, heights and isospectral sets of plane domains*, Math. Ann. **129** (1989), 293–362.

8. D. RAY and I. SINGER, *R-torsion and the Laplacian on Riemannian manifolds*, Adv. Math. **7** (1971), 145210.
9. S. ROSENBERG, *The Laplacian on a Riemannian manifold*, London Math. Soc. **31**, Cambridge University Press (1997).
10. J. ROWLETT, *Spectral geometry and asymptotically conic convergence*, Comm. Anal. Geom. **16** (2008), 735798.

Part III
Spectral and Scattering Theory

Relatively Isospectral Noncompact Surfaces

Pierre Albin, Clara Aldana, and Frédéric Rochon

I will report on work in progress with Clara Aldana and Frédéric Rochon extending the famous compactness result of Osgood, Phillips, and Sarnak from compact surfaces to non-compact surfaces.

Recall that two closed Riemannian surfaces (M, g) and (M', g') are said to be isospectral if their Laplacians Δ and Δ' have the same spectrum counted with multiplicity. It is well-known that the spectrum of a surface does not determine the metric, and so it is natural to ask how large the set of metrics with a given spectrum can be.

Richard Melrose [8] considered the case of planar domains and showed that the spectrum of the Laplacian (with Dirichlet boundary conditions) determines the geodesic curvature of the boundary of the domain to within a compact set. Indeed, he was able to show that the first k terms in the short-time expansion of the trace of the heat kernel (a spectral invariant) control the first k Sobolev norms of the geodesic curvature.

Melrose's result on these coefficients was extended to closed surfaces by Osgood et al. [6] (and eventually by Gilkey [2] to higher dimensional closed manifolds). In [6] the authors were also able to extend Melrose's compactness result. Indeed, they showed that starting with any sequence of isospectral metrics (M_i, g_i) on, say, closed surfaces there is a subsequence (M_{i_k}, g_{i_k}), a closed surface (M, g), and a sequence of diffeomorphisms $\phi_k : M \longrightarrow M_{i_k}$ such that

$$\phi_k^* g_{i_k} \longrightarrow g.$$

P. Albin (✉)
University of Illinois at Urbana-Champaign, Urbana, Illinois, USA
e-mail: palbin@illinois.edu

C. Aldana
Max-Planck-Institut für Gravitationsphysik, Potsdam, Germany
e-mail: clara.aldana@aei.mpg.de

F. Rochon
Université de Québec à Montréal, Montreal, Canada
e-mail: rochon.frederic@uqma.ca

D. Grieser et al. (eds.), *Microlocal Methods in Mathematical Physics and Global Analysis*, 87
Trends in Mathematics, DOI 10.1007/978-3-0348-0466-0_19, © Springer Basel 2013

This compactness result has been extended to non-compact surfaces in a couple of contexts. First in the setting of exterior domains, $\mathbb{R}^2 \setminus \mathcal{O}$, by Hassell and Zelditch [5]. Using that the metric is Euclidean near infinity and the resulting scattering theory to define the 'scattering phase' $s(\lambda)$, they say that two obstacles are *isophasal* if they have the same scattering phase, and they point out that this is analogous to using the counting function to express isospectrality of closed surfaces. They are able to prove a sequential compactness result for isophasal obstacles after using inversion with respect to the origin to work with bounded domains.

A second extension of the compactness result of Osgood-Phillips-Sarnak to non-compact surfaces was carried out by Borthwick and Perry [1]. They work in the context of non-compact surfaces whose ends are hyperbolic funnels and whose metrics coincide outside of some fixed compact set. The hyperbolicity of the metrics at infinity allows them to use a meromorphic continuation of the resolvent [3, 4, 7]. Two metrics are dubbed *isoresonant* if the poles of these meromorphic continuations coincide with multiplicity. Borthwick and Perry show that a sequence of isoresonant metrics on non-compact surfaces that coincide, and are hyperbolic funnels, outside a fixed compact set has a convergent subsequence in the same sense as in Osgood-Phillips-Sarnak.

Both of these results make strong assumptions about the structure of the metric near infinity. In the latter the ends are hyperbolic funnels while in the former the ends are Euclidean. Our aim is to prove a compactness result for non-compact surfaces with only weak assumptions about the behavior of the metrics near infinity, although we do require that the metrics coincide outside of a compact set. An important first step is to extend the notion of isospectral metrics.

We say that two Riemannian spaces (M_1, g_1), (M_2, g_2) coincide cocompactly if there are compact subsets

$$K_1 \subseteq M_1, \quad K_2 \subseteq M_2$$

and an isometry $(M_1 \setminus K_1, g_1) \longrightarrow (M_2 \setminus K_2, g_2)$. We say that (M_1, g_1) and (M_2, g_2) are relatively isospectral if they coincide cocompactly and

$$\mathrm{Tr}(e^{-t\Delta_{g_1}} - e^{-t\Delta_{g_2}}) = 0 \text{ for all } t > 0.$$

We are implicitly assuming that $e^{-t\Delta_{g_1}} - e^{-t\Delta_{g_2}}$ is trace-class, however this is automatically true if (M_1, g_1) is complete and is also true in some singular contexts. Notice that this condition extends the notion of isospectrality on closed manifolds. Indeed, if (M_i, g_i) are compact and have the same trace of the heat kernel for all times, then they have the same spectrum with multiplicity as one can see from the uniqueness of the Laplace transform of a measure.

Our main result is a sequential compactness theorem for relatively isospectral metrics on non-compact surfaces with some mild assumptions. Details will be forthcoming.

References

1. David Borthwick and Peter A. Perry. Inverse scattering results for metrics hyperbolic near infinity. preprint (2009) to appear in *J. Geom. Anal.*
2. Peter B. Gilkey. Leading terms in the asymptotics of the heat equation. In *Geometry of random motion (Ithaca, N.Y., 1987)*, volume 73 of *Contemp. Math.*, pages 79–85. Amer. Math. Soc., Providence, RI, 1988.
3. Colin Guillarmou. Meromorphic properties of the resolvent on asymptotically hyperbolic manifolds. *Duke Math. J.*, 129(1):1–37, 2005.
4. Laurent Guillopé and Maciej Zworski. Upper bounds on the number of resonances for noncompact Riemann surfaces. *J. Funct. Anal.*, 129(2):364–389, 1995.
5. Andrew Hassell and Steve Zelditch. Determinants of Laplacians in exterior domains. *Internat. Math. Res. Notices*, (18):971–1004, 1999.
6. Brian Osgood, Ralph Phillips, and Peter Sarnak. Compact isospectral sets of surfaces. *J. Funct. Anal.*, 80(1):212–234, 1988.
7. Rafe Mazzeo and Richard B. Melrose. Meromorphic extension of the resolvent on complete spaces with with asymptotically negative curvature. *J. Funct. Anal.*, pages 260–310, 1987.
8. Richard B. Melrose. Isospectral sets of drumheads are compact in C^∞. Unpublished preprint from MSRI, 048-83. Available online at http://math.mit.edu/~rbm/paper.html, 1983.

Microlocal Analysis of Scattering Data for Nested Conormal Potentials

Suresh Eswarathasan

1   Statement of the Problem and Results

Consider the potential scattering problem for the wave equation:

$$(\partial_t^2 - \triangle + q)u = 0 \text{ in } \mathbb{R}^n \times \mathbb{R}$$

$$u = \delta(t - x \cdot \omega) \text{ for } t << -\rho, \tag{1}$$

where q is a compactly supported potential and $\omega \in S^{n-1}$ is fixed. Here ρ is the value such that $supp(q) \subset \{|x| \le \rho\}$.

The scattering map Φ which sends q to α_q is nonlinear and overdetermined and there has been much interest in the inverse problem of determining q from α_q. Since α_q is overdetermined, it is naturally also of interest to try to reconstruct q from the restriction of α_q to various submanifolds of $\mathbb{R} \times S^{n-1} \times S^{n-1}$. In this work, we are interested in the inverse problem of determining some information about q from restrictions of α_q to some lower-dimensional set.

Let us review a few of the important inverse scattering results involving fixed angle and backscattering data.

1. *Fixed angle scattering*: Set $\theta = \theta_0 \in S^{n-1}$. Stefanov [9] proves uniqueness of the potentials under a smallness assumption and Ruiz [8] shows that the Born approximation determines a "close" approximation of $q \in H^s(\mathbb{R}^n)$ for $n = 2$ and 3.

S. Eswarathasan (✉)
Department of Mathematics, Hylan Building, University of Rochester, Rochester, NY 14623, USA

Current address: Centre de recherches mathématiques, Université de Montréal, P.O. Box 6128, Centre-ville Station, Montréal (Québec) H3C 3J7, Canada
e-mail: seswarathasan@gmail.com

D. Grieser et al. (eds.), *Microlocal Methods in Mathematical Physics and Global Analysis*, Trends in Mathematics, DOI 10.1007/978-3-0348-0466-0_20, © Springer Basel 2013

2. *Backscattering*: Set $\theta = -\omega \in S^{n-1}$. Uniqueness under a smallness assumption is obtained by Lagergen [6] for $n = 3$ and recovery of singularities for $n = 2$ by Ola et. al [7]. Generic uniqueness is proven in Eskin and Ralston [1].

The above referenced papers contain further citations of articles with significant results in inverse scattering. A highly pursued question within the inverse problems community is the following:

Open problem: For what spaces of potentials q and partial sets of scattering data do we have uniqueness and/or reconstruction? That is, for which $\mathbb{D} \subset \mathbb{R} \times S^{n-1} \times S^{n-1}$ does $\alpha_q|_{\mathbb{D}}$ determine q? For these, can one constructively determine (reconstruct) q from $\alpha_q|_{\mathbb{D}}$?

The class of q's to be considered in (1) are those that have singularities conormal to a nested pair of submanifolds $S_2 \subset S_1$ of $\mathbb{R}^n$, denoted by $I^{\mu_1,\mu_2}(S_1, S_2)$ [2], with S_1 and S_2 intersecting in codimension d, and μ_1 and μ_2 being the singularity orders; these are a subset of the paired Lagrangian distributions introduced in [4,5]. The inverse problem that is solved consists of determining these submanifolds and the principal symbol of q, which is enough to determine the singularities of q, from the leading singularities of the backscattering $\alpha_q|_{\mathbb{B}}$, where $\mathbb{B} = \{\theta = -\omega\} \subset \mathbb{R} \times S^{n-1} \times S^{n-1}$. In addition, we treat similar determined sets of scattering data.

It is shown that α_q is, away from ω's that are tangent to either of the submanifolds, the sum of a paired Lagrangian distribution associated to two cleanly intersecting reflected Lagrangians, two reflected Lagrangian distributions, and a single peak Lagrangian distribution, modulo Sobolev errors. Although the strongest singularity lies on the peak Lagrangian, which is consistent with the physics literature, we show that it is the restriction of the reflected Lagrangians and their points of intersection to various submanifolds of scattering data in $\mathbb{R} \times S^{n-1} \times S^{n-1}$ that determine the singularities of q. The precise theorem is the following:

Theorem 1. *Let $S_2 \subset S_1 \subset \mathbb{R}^n$ be smooth nested submanifolds of codimension $d_1 + d_2$ and d_1, respectively. Assume that q is compactly supported and is conormal to the nested pair (S_1, S_2) of orders M_1 and M_2. Furthermore, suppose that*

$$M_2 > -d_2 \text{ and } M_1 < -d_1 - \frac{d_2}{2} + 1 \text{ or}$$

$$M_2 \leq -d_2 \text{ and } M_1 < -d_1 + 1, \text{ for}$$

$$M_1 + \frac{M_2}{2} < \inf\{-\frac{n-2}{n}(d_1 + d_2), -d_1 - d_2 + 1\} \text{ if } n \geq 5,$$

and

$$M_1 + \frac{M_2}{2} < \inf\{-\frac{d_1 + d_2}{2}, -d_1 - d_2 + 1\} \text{ if } n = 3 \text{ and } 4.$$

Then S_1, S_2, and the principal symbol of q are determined by the singularities of α_q restricted to the backscattering surface $\{\theta = -\omega\} \subset \mathbb{R} \times S^{n-1} \times S^{n-1}$.

In fact, we prove a stronger result that shows that for a given submanifold of scattering data with certain geometrical properties, the inverse problem can still be solved. See [2] for a more detailed statement.

Potentials that are conormal to a single submanifold are dealt with by Greenleaf and Uhlmann [3]; we closely follow the time-dependent approach taken in [3] and generalize the results to these more singular q. Using Proposition 3.7 in [4] involving the intersection of classes of paired Lagrangians over all orders M_2, it follows that Theorem 1 covers the main result of [3]. The potentials under consideration are allowed to blow up, i.e. there are no size restrictions on q, in contrast to those made in, e.g. [9].

A significant difference between this work and that of [3] is the new, more complicated geometry that arises when using an approximation method, the understanding of multiplication by q and the parametrix to $\Box$ on Sobolev spaces and other classes of distributions, and the appearance of distributions that are associated to cleanly intersecting triples and quadruples of Lagrangians.

We note that even for a arbitrary Lagrangian distribution u, calculating the blowup rates that assist in finding which L^p space u belongs to is difficult without some additional assumptions on the Lagrangian. Hence, from the viewpoint of the Lax-Phillips scattering theory, assuming that u is in some conormal category is a reasonable restriction.

2   Brief Outline of the Proof

The goal of recovering singularities lies in understanding an approximation to the scattering kernel α_q. We start by analyzing the forward problem and the Born series,

$$\sum_i (-1)^i (\Box^{-1} M_q)^i (\delta(t - x \cdot \omega)), \tag{2}$$

which is an asymptotic expansion of the solution u to (1), where $\Box^{-1}$ is the parametrix to the wave equation and M_q is multiplication by the nested conormal distribution q. A "good" description of the singularties of (2) will eventually lead, after application of certain elliptic Fourier integral operators, to a precise enough approximation of $\alpha_{|\mathbb{B}}$ that will solve our inverse problem.

Due to the proliferation of new wavefront set that occurs at each term of the Born series, we concentrate on $u_0 + u_1$, referred to by physicists as the Born approximation. As a result of the multiplication of a paired Lagrangian distribution against a single Lagrangian distribution, we develop a natural class of distributions that are conormal to a nested triple of submanifolds. Also, the action of $\Box^{-1}$ on the paired Lagrangian and nested triple spaces is studied in order to get the most refined description of the singularities of the Born approximation that is possible.

After understanding the strength and location of the singularities of $u_0 + u_1$, we need to show the remaining terms of the Born series are smoother in Sobolev regularity than of the previous terms. This is equivalent to showing that our error terms do not "interfere" with the measurements. We take this approach because of the proliferating wavefront set mentioned earlier. The Sobolev mapping properties of $\Box^{-1}$ are well understood, but those of M_q are not. The absence of a composition calculus for operators of this type, specifically those whose Schwarz kernels have singularities on intersecting triples of Lagrangians, brings us to use a parabolic cut-off argument of Melrose and along with Littlewood-Paley techniques. The orders that appear in the statement of Theorem 1 are derived after requiring that the composition $\Box^{-1} M_q$ gain derivatives.

Once we have shown the required smoothness, an application of a Radon-type transform and a pullback operator to this error, following the Lax-Phillips formulation of the scattering theory, shows that we can formulate and use an approximate scattering kernel that displays the same singularities as those of the exact scattering kernel. Geometrical computations, an application of the restriction operator to the backscattering (also an elliptic Fourier integral operator), and the symbol calculus in [4] essentially finish our problem. A fairly straightforward geometrical lemma settles the result for more general determined sets of scattering data.

Acknowledgements The author would like to thank his PhD advisor, Allan T. Greenleaf, under which this research was done.

References

1. G. Eskin and J. Ralston, *The inverse backscattering problem in three dimensions*, Comm. Math. Phys.. 124: 169–215, 1989.
2. Suresh Eswarathasan, *Microlocal analysis of scattering data for nested conormal potentials*, preprint (2011), available at http://arxiv.org/abs/1103.6015.
3. Allan Greenleaf and Gunther Uhlmann, *Recovering singularities of a potential from singularities of scattering data*, Comm. Math. Phys.. 157: 549–572, 1993.
4. Victor Guillemin and Gunther Uhlmann, *Oscillatory integrals and singular symbols*, Duke Math. J.. 48: 251–267, 1981.
5. Richard Melrose and Gunther Uhlmann, *Lagrangian intersection and the Cauchy problem*, Comm. Pure Appl. Math.. 32: 483–519, 1979.
6. Robert Per Lagergen, *The back-scattering problem in three dimensions*, ProQuest LLC, Ann Arbor, MI, 2011, Thesis (Ph.D.) - Lund University (Sweden).
7. Petri Ola, Lassi Päiv ainta and Valeri Serov, *Recovering singularities from backscattering in two dimensions*, Comm. Partial Differential Equations. 26: 697–715, 2001.
8. Alberto Ruiz, *Recovery of the singularities of a potential from fixed angle scattering data*, Comm. Partial Differential Equations. 26: 1721–1738, 2001.
9. Plamen Stefanov, *Generic uniqueness for two inverse problems in potential scattering*, Comm. Partial Differential Equations. 17: 55–68, 1992.

Equidistribution of Eisenstein Series for Convex Co-compact Hyperbolic Manifolds

Colin Guillarmou and Frédéric Naud

The quantum ergodic theorem, due to Schnirelman [4], Colin de Verdière [1] and Zelditch [5], says that on any compact Riemannian manifold X whose geodesic flow is ergodic, one can find a full density sequence $\lambda_j \to +\infty$ of eigenvalues of the Laplacian Δ_X such that the corresponding normalized eigenfunctions ψ_j are equidistributed i.e. for all $f \in L^2(X)$, we have

$$\lim_{j \to +\infty} \int_X f(z)|\psi_j(z)|^2 dv(z) = \int_X f(z)dv(z),$$

where dv is the normalized volume measure. For non-compact manifolds, there can be continuous spectrum and the quantum ergodic theorem does not really make sense in general. However, for hyperbolic surfaces of finite volume and in particular arithmetic cases, Zelditch [6], Luo-Sarnak [3] prove a related statement involving the generalized eigenfunctions (also known as Eisenstein series). Let us recall their results. Let $X = \Gamma \backslash \mathbb{H}^2$ be a finite area surface where Γ is a non co-compact co-finite Fuchsian group. The non compact ends of X are cusps related to fixed points c_j in $\partial \mathbb{H}^2$ of parabolic elements in Γ. The spectrum of the Laplacian Δ_X has a discrete part which corresponds to $L^2(X)$-eigenfunctions and may be infinite and the absolutely continuous part $[1/4, +\infty)$ which is parametrized ($t \in \mathbb{R}$) by the finite set of Eisenstein series $E_X(1/2 + it; z, j)$ related to each cusp c_j. The Eisenstein series $E_X(1/2 + it; z, j)$ are smooth non-$L^2(X)$ eigenfunctions

C. Guillarmou (✉)
DMA, U.M.R. 8553 CNRS, Ecole Normale Supérieure, 45 rue d'Ulm,
F 75230 Paris cedex 05, France
e-mail: cguillar@dma.ens.fr

F. Naud
Laboratoire d'Analyse non linéaire et Géométrie, Université d'Avignon,
33 rue Louis Pasteur 84000 Avignon, France
e-mail: frederic.naud@univ-avignon.fr

D. Grieser et al. (eds.), *Microlocal Methods in Mathematical Physics and Global Analysis*, 95
Trends in Mathematics, DOI 10.1007/978-3-0348-0466-0_21, © Springer Basel 2013

$$\Delta_X E_X(1/2 + it; z, j) = (1/4 + t^2) E_X(1/2 + it; z, j).$$

For all $t \in \mathbb{R}$, define the density μ_t by

$$\int_X a(z) d\mu_t(z) := \sum_j \int_X a(z) |E_X(1/2 + it; z, j)|^2 dv(z),$$

where $a \in C_0^\infty(X)$. In the case with only finitely many eigenvalues then Zelditch's equidistribution result is as follows: for $a \in C_0^\infty(X)$,

$$\frac{1}{s(T)} \int_{-T}^T \left| \int_X a d\mu_t - \partial_t s(t) \int_X a \, dv \right| dt \to 0 \text{ as } T \to \infty$$

where $s(t)$ is the scattering phase appearing as a sort of regularization of Eisenstein series due to the fact that the Weyl law involves the continuous spectrum. On the other hand, for the modular surface $X = \mathrm{PSL}_2(\mathbb{Z}) \backslash \mathbb{H}^2$, Luo and Sarnak [3] showed that as $t \to +\infty$,

$$\int_X a d\mu_t = \frac{48}{\pi} \log(t) \int_X a \, dv + o(\log(t)),$$

which is a much stronger statement obtained via sharp estimates on certain L-functions.

We report here some recent result of [2], where we studied the case of *infinite volume hyperbolic manifolds* without cusps, more precisely convex co-compacts quotients $X = \Gamma \backslash \mathbb{H}^{n+1}$ of the hyperbolic space. A discrete group of orientation preserving isometries of $\mathbb{H}^{n+1}$ is said to be convex co-compact if it admits a polygonal, finite sided fundamental domain whose closure does not intersect the limit set of Γ. The limit set Λ_Γ and the set of discontinuity Ω_Γ are defined by

$$\Lambda_\Gamma := \overline{\Gamma.o} \cap S^n, \quad \Omega_\Gamma := S^n \setminus \Lambda_\Gamma,$$

where $o \in \mathbb{H}^{n+1}$ is any point in $\mathbb{H}^{n+1}$. The quotient space $X = \Gamma \backslash \mathbb{H}^{n+1}$ has 'funnel type' ends and is the interior of a compact manifold with boundary $\overline{X} := \Gamma \backslash (\mathbb{H}^{n+1} \cup \Omega_\Gamma)$, the action of Γ on $(\mathbb{H}^{n+1} \cup \Omega_\Gamma)$ being free and totally discontinuous. By a result of Patterson and Sullivan, the Hausdorff dimension of Λ_Γ

$$\delta_\Gamma := \dim_{\mathrm{Haus}}(\Lambda_\Gamma)$$

is also the exponent of convergence of the Poincaré series, i.e. for all $m, m' \in \mathbb{H}^{n+1}$ and $s > 0$,

$$\sum_{\gamma \in \Gamma} e^{-sd(\gamma m, m')} < \infty \iff s > \delta_\Gamma, \tag{1}$$

where $d(m, m')$ denotes the hyperbolic distance.

In that case the spectrum of Δ_X consists of the absolutely continuous spectrum $[n^2/4, +\infty)$ and a (possibly empty) finite set of eigenvalues in $(0, n^2/4)$. The Eisenstein functions are defined using the ball model of $\mathbb{H}^{n+1}$ to be the automorphic functions of $m \in \mathbb{H}^{n+1}$ given by

$$E_X(s; m, \xi) = \sum_{\gamma \in \Gamma} \left(\frac{1 - |\gamma m|^2}{4|\gamma m - \xi|^2} \right)^s, \quad \xi \in \Omega_\Gamma,$$

which are absolutely convergent for $\mathrm{Re}(s) > \delta_\Gamma$ and extend meromorphically to $s \in \mathbb{C}$. The Eisenstein series are non-$L^2(X)$ eigenfunctions of the Laplacian with eigenvalue $s(n - s)$ on $\mathrm{Re}(s) = n/2$. We show the following

Theorem 1. *Let $X = \Gamma \backslash \mathbb{H}^{n+1}$ be a convex co-compact quotient with $\delta_\Gamma < n/2$. Let $a \in C_0^\infty(X)$ and let $E_X(s; \cdot, \xi)$ be an Eisenstein series as above with a given point $\xi \in \partial \overline{X}$ at infinity. Then we have as $t \to +\infty$,*

$$\int_X a(m) \left| E_X\left(\frac{n}{2} + it; m, \xi\right) \right|^2 dv(m) = \int_X a(m) E_X(n; m, \xi) dv(m) + \mathcal{O}(t^{2\delta_\Gamma - n}).$$

The limit measure on X is given by the harmonic density $E_X(n; m, \xi)$ whose boundary limit is the Dirac mass at $\xi \in \partial \overline{X}$. A microlocal extension of this theorem is also proved. We first need to introduce some adequate notations. Fix any $\xi \in \partial \overline{X}$. Let $\mathcal{L}_\xi^\Gamma$ defined by

$$\mathcal{L}_\xi^\Gamma := \overline{\cup_{\gamma \in \Gamma} \mathcal{L}_{\gamma \xi}} \subset S^* X,$$

where $\mathcal{L}_{\gamma \xi}$ are stable Lagrangian submanifolds of the unit cotangent bundle $S^* X$: the Lagrangian manifold $\mathcal{L}_{\gamma \xi}$ is defined to be the projection on $\Gamma \backslash S^* \mathbb{H}^{n+1}$ of

$$\{(m, \nu_{\gamma \xi}(m)) \in S^* \mathbb{H}^{n+1}; m \in \mathbb{H}^{n+1}\},$$

where $\nu_{\gamma \xi}(m)$ is the unit (co)vector tangent to the geodesic starting at m and pointing toward $\gamma \xi \in S^n$. The set $\mathcal{L}_\xi^\Gamma$ "fibers" over X, and the fiber over a point $m \in X$ corresponds to the closure of the set of directions $v \in S^* X$ such that the geodesic starting at m with directions v converges to $\xi \in \partial \overline{X}$ as $t \to +\infty$. Since the closure of the orbit $\Gamma.\xi$ satisfies $\overline{\Gamma.\xi} \supset \Lambda_\Gamma$, $\mathcal{L}_\xi^\Gamma$ contains the forward trapped set

$$\mathcal{T}_+ := \{(m, v) \in S^* X \ : \ g_t(m, v) \text{ remains bounded as } t \to +\infty\},$$

where $g_t : S^* X \to S^* X$ is the geodesic flow. The Hausdorff dimension of $\mathcal{L}_\xi^\Gamma$ is $n + \delta_\Gamma + 1$ and satisfies $n + 1 < \delta_\Gamma + n + 1 < 2n + 1$ if Γ is non elementary.
    Our phase-space statement is the following

Theorem 2. *Let A be a compactly supported 0-th order pseudodifferential operator with principal symbol $a \in C_0^\infty(X, T^* X)$, then as $t \to +\infty$*

$$\left\langle AE_X(\frac{n}{2}+it;\cdot,\xi), E_X(\frac{n}{2}+it;\cdot,\xi)\right\rangle_{L^2(X)} = \int_{S^*X} a\, d\mu_\xi + \mathcal{O}(t^{-\min(1,n-2\delta_\Gamma)})$$

where μ_ξ is a g_t-invariant measure supported on the fractal subset $\mathcal{L}_\xi^\Gamma \subset S^*X$.

Notice that the fractal behaviour of the semi-classical limit μ_ξ can only be observed at the microlocal level. By averaging over the boundary with respect to the volume measure induced by S^n on Ω_Γ, we obtain as $t \to +\infty$

$$\int_{\partial\overline{X}}\int_X a(m)\left|E_X(\frac{n}{2}+it;m,\xi)\right|^2 dv(m)d\xi = \text{vol}(S^n)\int_X a(m)dv(m) + \mathcal{O}(t^{2\delta_\Gamma-n})$$

(2)

and

$$\int_{\partial\overline{X}}\left\langle AE_X(\frac{n}{2}+it;\cdot,\xi), E_X(\frac{n}{2}+it;\cdot,\xi)\right\rangle_{L^2(X)} d\xi = \int_{S^*X} a\, d\mu + \mathcal{O}(t^{-\min(n-2\delta_\Gamma,1)})$$

where μ denotes the Liouville measure. This is the perfect analog of the previously known results for the modular surface (actually with a remainder in our case).

References

1. Y. Colin de Verdière, *Ergodicité et fonctions propres du Laplacien*, Comm. Math. Phys., **102** (1985), 497–502.
2. C. Guillarmou, F. Naud, *Equidistribution of Eisenstein series on convex co-compact hyperbolic manifolds*, arXiv:1107.2655
3. W. Luo, P. Sarnak, *Quantum ergodicity of eigenfunctions on $PSL_2(\mathbb{Z})/\mathbb{H}^2$*, Publications Mathématiques de l'IHES, **81** (1995), 207–237.
4. A. I. Schnirelman, *Ergodic properties of eigenfunctions*, Usp. Math. Nauk., **29** (1974), 181–182.
5. S. Zelditch, *Uniform distribution of eigenfunctions on compact hyperbolic surfaces*, Duke Math. J. **55** (1987), 919–941.
6. S. Zelditch, *Mean Lindelöf hypothesis and Equidistribution of Cusps forms and Eisenstein series*, Journal of Functional Analysis **97** (1991), 1–49.

Lower Bounds for the Counting Function of an Integral Operator

Yuri Safarov

1 Motivation

Let $\mathbf{a}[\cdot]$ be a closed quadratic form defined on a subspace H^1 of a Hilbert space H, and let H_0^1 be an $\mathbf{a}$-closed subspace of H^1 which is dense in H. Consider the self-adjoint operators A_N and A_N generated by the form $\mathbf{a}[\cdot]$ with domains H^1 and H_0^1 respectively, and denote by $N_N(\lambda)$ and $N_D(\lambda)$ their left continuous counting functions.

Example 1. If Ω is a domain in $\mathbb{R}^n$, $n \geq 2$, $H := L_2(\Omega)$, H^1, H_0^1 are the Sobolev spaces and $\mathbf{a}[u] := \int_\Omega |\nabla u(x)|^2 \, \mathrm{d}x$ then A_D and A_N are the Dirichlet and Neumann Laplacians on Ω. More generally, if γ is a measurable function on the boundary $\partial\Omega$ and $\mathbf{a}[u] := \int_\Omega |\nabla u(x)|^2 \, \mathrm{d}x + \int_{\partial\Omega} \gamma(x') |u(x')|^2 \, \mathrm{d}x'$ then A_N is the Laplacian with Robin boundary condition.

Theorem 1. *If Ω in Example 1 is a bounded domain with smooth boundary and $\gamma \equiv 0$ then $N_N(\lambda) = N_D(\lambda) + n_D(\lambda) + g_-(\lambda)$, where $n_D(\lambda)$ is the multiplicity of λ as an eigenvalue of A_D and $g_-(\lambda)$ is the number of negative eigenvalues of the Dirichlet-to-Neumann (D-N) operator in the subspace $\{u \in H^1 : -\Delta u = \lambda u\}$* (see [2]).

Theorem 2. *The same is true for a bounded domain Ω with smooth boundary in a Riemannian manifold* (see [3]).

Remark 1. In [1], N. Filonov noticed that, under the conditions of Theorem 1, $N_N(\lambda) \geq N_D(\lambda) + n_D(\lambda) + g_-(\lambda)$ for all bounded domains Ω.

Y. Safarov (✉)
Department of Mathematics, King's College London, London WC2R 2LS, UK
e-mail: yuri.safarov@kcl.ac.uk

D. Grieser et al. (eds.), *Microlocal Methods in Mathematical Physics and Global Analysis,*
Trends in Mathematics, DOI 10.1007/978-3-0348-0466-0_22, © Springer Basel 2013

Let A be the unbounded (non-self-adjoint) operator such that $Au = f$ if and only if $\mathbf{a}[u,v] = (Au,v)_H$ for all $v \in H_0^1$ (wee assume that $u \in \mathcal{D}(A)$ if such a function f exists). In the abstract setting, the D-N operator is understood as the self-adjoint operator generated by the restriction of the quadratic form $\mathbf{a}[u] - \lambda \|u\|_H^2$ to the subspace $G_\lambda := \{u \in \mathcal{D}(A) : Au = \lambda u\}$.

Example 2. If $\mathbf{a}[\cdot]$ is a differential form on a domain Ω then A is the differential operator on Ω without boundary condition on Ω obtained by integration by parts in $\mathbf{a}[\cdot]$. The classical D-N operator on $\partial\Omega$ is obtained by integration by parts in the form $\mathbf{a}[u] - \lambda \|u\|_H^2$.

Theorem 3. *Theorem 1 holds in the abstract setting, no conditions are needed* (see [4]).

Under the conditions of Theorem 1, the functions $e_\xi(x) := e^{ix \cdot \xi}$ with $\{\xi \in \mathbb{R} : |\xi|^2 = \lambda > 0\}$ belong to G_λ. Clearly, $\mathbf{a}[e_\xi] = \lambda \|e_\xi\|_H^2$, which implies that $g_-(\lambda) \geq 1$ and, consequently, $N_N(\lambda) - N_D(\lambda) \geq 1$. This shows that there are at least $k+1$ Neumann eigenvalues which are smaller than the kth Dirichlet eigenvalue.

In view of the above results, the same is true in the abstract setting, provided that there exists $e_\lambda \in G_\lambda$ such that $\mathbf{a}[e_\lambda] \leq \lambda \|e_\lambda\|^2$ and e_λ is not an eigenfunction.

Note that e_ξ form an infinite dimensional subset of $H = L_2(\Omega)$. But this subset contains only one dimensional linear subspaces, so that we can only say that $g_-(\lambda) \geq 1$.

Question: is it possible to construct a higher dimensional "negative" subspace of the quadratic form $\mathbf{a}[e_\lambda] - \lambda \|e_\lambda\|^2$, using the functions e_ξ ?

2 Reduction to the Integral Operator

Let M be a metric space with a locally finite Borel measure v, and let $\{e_\xi\}_{\xi \in M}$ is a continuous family of solutions to the equation $Au = \lambda u$. Define

$$\mathcal{K}(\xi, \eta) := \mathbf{a}[e_\xi, e_\eta] - \lambda\,(e_\xi, e_\eta)_H$$

and consider the self-adjoint integral operator $K : u \mapsto \int \mathcal{K}(\xi, \eta)\, u(\eta)\, dv(\eta)$ in the space $L_2(M, dv)$. If $f_u(\xi) := (u, e_\xi)_{L_2(M,dv)}$ then $\mathbf{a}[f_u] - \lambda \|f_u\|^2 = (Ku, u)_{L_2(M,dv)}$. This implies

Lemma 1. *Assume that $f_u \not\equiv 0$ for all nonzero functions $u \in L_2(M, dv)$. Then $g_-(\lambda)$ is estimated from below by the number of negative eigenvalues of K.*

Remark 2. We can choose the measure v on M as we wish. In particular, if v is the sum of δ-measures located at some points $\xi_k \in M$ then K coincides with the restriction of the D-N operator to the subspace spanned by the functions e_{ξ_k}.

Example 3. Let A_D and A_N be the Dirichlet and Robin Laplacians on a domain $\Omega \subset \mathbb{R}^n$ (as in Example 1). Take $M = \mathbb{S}_\lambda^{n-1} := \{\xi \in \mathbb{R}^n : |\xi| = \lambda\}$ and $e_\xi = e^{ix \cdot \xi}$. Then $\mathcal{K}(\xi, \eta) = -\frac{1}{2}|\eta - \xi|^2 \hat{\chi}_\Omega(\eta - \xi) + \hat{\gamma}_{\partial\Omega}(\eta - \xi)$ where $\hat{\chi}_\Omega(\theta) = \int_\Omega e^{-ix \cdot \theta}\, dx$ and $\gamma_{\partial\Omega}(\theta) = \int_{\partial\Omega} e^{ix \cdot \theta} \gamma(x')\, dx'$.

Assume that $\frac{1}{2}|\eta - \xi|^2 \hat{\chi}_\Omega(\eta - \xi) = \hat{\gamma}_{\partial\Omega}(\eta - \xi)$ for some $\theta \neq 0$. Then $\mathcal{K}(\xi, \xi) = \mathcal{K}(\xi, \eta) = \mathcal{K}(\eta, \xi) = \mathcal{K}(\eta, \eta) = 0$ for all $\xi, \eta \in \mathbb{S}_\lambda^{n-1}$ such that $\xi - \eta = \theta$. It follows that $g_-(\lambda) \geq 2$ for all $\lambda \geq |\theta|/2$.

3 Main Theorem

Let $\kappa(\xi, \eta)$ be the minimal eigenvalue of the matrix $\begin{pmatrix} \mathcal{K}(\xi, \xi) & \mathcal{K}(\xi, \eta) \\ \mathcal{K}(\eta, \xi) & \mathcal{K}(\eta, \eta) \end{pmatrix}$. Define $\Sigma_t := \{(\xi, \eta) \in M \times M : \kappa(\xi, \eta) < t\}$, $t \leq 0$. Clearly, Σ_t is an open symmetric subset of $M \times M$ and $\kappa(\xi, \eta) = \kappa(\eta, \xi)$. Let M_t be the projection of Σ_t onto M. If μ is a symmetric Borel measure on Σ_t, we shall denote by μ' its marginal, that is, the measure on M_t such that $\mu'(S) := \mu(\{(\xi, \eta) \in \Sigma_t : \xi \in S\})$ for all measurable $S \subset M_t$.

Theorem 4. *Let $\mathcal{N}(K; t)$ be the dimension of the eigenspace of K corresponding to the interval $(-\infty, t)$, where $t \leq 0$. If the set Σ_t is not empty then $\mathcal{N}(K, t) \geq \frac{1}{2} + \frac{1}{16} C_t(\mu)$ for all symmetric Borel measures μ on Σ_t, where*

$$C_t(\mu) := \frac{\left(\int_{\Sigma_t} (t - \kappa(\xi, \eta))\, d\mu(\xi, \eta)\right)^2}{\int_{M_t} \int_{M_t} |\mathcal{K}(\xi, \eta)|^2\, d\mu'(\xi)\, d\mu'(\eta)}.$$

Since $C_t(\mu) \geq 0$, Theorem 4 immediately implies that $\mathcal{N}(K, t) \geq 1$ whenever $\Sigma_t \neq \emptyset$. Applying the theorem with $\mu(\xi, \eta) = \mu'(\eta)\, \delta(\xi - \eta)$, we obtain

Corollary 1. *Let $M_- := \{\xi \in M : \mathcal{K}(\xi, \xi) < 0\} \neq \emptyset$. Then*

$$\mathcal{N}(K, 0) \geq \frac{1}{2} + \frac{\left(\int_{M_-} \mathcal{K}(\xi, \xi)\, d\mu'(\xi)\right)^2}{4 \int_{M_-} \int_{M_-} |\mathcal{K}(\xi, \eta)|^2\, d\mu'(\xi)\, d\mu'(\eta)}$$

for all Borel measures μ' on M_-.

A possible strategy of optimizing the choice of μ in Theorem 4 is to fix the marginal μ' and to minimize $\int(t - \kappa(\xi, \eta)\, d\mu(\xi, \eta)$ over the set of symmetric measures μ with the fixed marginal. The problem of minimizing the integral of the form $\int f(\xi, \eta)\, d\mu(\xi, \eta)$ over the set of measures with fixed marginals is known as Kantorovich's problem. It has been solved for some special functions $f(\xi, \eta)$.

References

1. N. Filonov. *On an inequality between Dirichlet and Neumann eigenvalues for the Laplace operator*, Algebra Anal. **16**, No.2 (2004), 172–176 (Russian). English translation in St. Petersbg. Math. J. **16**, No. 2 (2005), 413–416.
2. L. Friedlander. *Some inequalities between Dirichlet and Neumann eigenvalues*, Arch. Ration. Mech. Anal. **116** (1991), 153–160.
3. R. Mazzeo. *Remarks on a paper of Friedlander concerning inequalities between Neumann and Dirichlet eigenvalues,* Int. Math. Res. Not., No. 4 (1991), 41–48.
4. Y. Safarov, *On the comparison of the Dirichlet and Neumann counting functions*, AMS Translations (2), Advances in Mathematical Sciences, vol. 225 (2008), 191–204.

The Identification Problems in SPECT: Uniqueness, Non-uniqueness and Stability

Plamen Stefanov

We study an inverse problem arising in Single Photon Emission Computerized Tomography (SPECT): recover an unknown source distribution with a unknown attenuation. The mathematical model is the attenuated X-ray transform

$$X_a f(x, \theta) = \int e^{-Ba(x+t\theta,\theta)} f(x + t\theta)\, \mathfrak{t}, \quad x \in \mathbf{R}^2, \ \theta \in S^1, \tag{1}$$

in the plane with a source f and an attenuation a that we want to recover. We denote by

$$Ba(x, \theta) = \int_0^\infty a(x + t\theta)\, dt \tag{2}$$

the "beam transform" of a, usually denoted by Da. The functions a and f are assumed to be compactly supported. We analyze whether one can recover both a and f, and if so; whether this can be done in a stable way.

The linearization $\delta X_{a,f}$ of $X_a f$ with respect to (a, f) turns out to be a sum of two weighted X-ray transforms:

$$\delta X_{a,f}(\delta a, \delta f) = I_w \delta a + X_a \delta f,$$

where

$$I_w f(x, \theta) = \int w(x + t\theta, \theta) f(x + t\theta)\, dt, \quad x \in \mathbf{R}^2, \ \theta \in S^1,$$

P. Stefanov (✉)
Department of Mathematics, Purdue University, west lafayette, IN, USA
e-mail: stefanov@math.purdue.edu

D. Grieser et al. (eds.), *Microlocal Methods in Mathematical Physics and Global Analysis*, 103
Trends in Mathematics, DOI 10.1007/978-3-0348-0466-0_23, © Springer Basel 2013

and

$$w = -e^{-Ba}u,$$

with

$$u(x, \theta) = \int_{-\infty}^{0} e^{-\int_{t}^{0} a(x+s\theta)\,ds} f(x + t\theta)\,dt.$$

This motivates the study of the inversion of the more general transform $I_{w_1}g_1 + I_{w_2}g_2$.

The data $\delta X_{a,f}(\delta a, \delta f)$ contains an integral over each line twice—once in each direction. On the other hand, the integral over each such line (and some neighborhood of it) contains information about singularities conormal to it. So microlocally, we have a 2×2 system. To get a pseudo-differential system, instead of an FIO one, we take the Fourier transform with respect to the initial point $z \in \theta^{\perp}$ of each line, for each direction θ. The determinant of the so transformed $\delta X_{a,f}$, up to elliptic factors, is

$$p_0(x, \xi) = u(x, \theta) - u(x, -\theta)|_{\theta = \xi^{\perp}/|\xi|}.$$

Since p_0 is an odd function of θ, it has zeros over any point x. Therefore, elliptic methods would not work. The Hamiltonian flow of p_0 then plays an important role. We call the projections of the zero bicharacteristics of p_0 to the x-space *rays*. By the Duistermaat-Hörmaner propagation of singularities theorem, on any bicharacteristics, either each point is a singularity, or none is. If we a priori know that δa, δf are supported in a non-trapping set K (no ray lies entirely in K), then recovery of the singularities of δa, δf is possible with a loss of one derivative. Then we also have an a priori regularity estimate. A more careful analysis actually reveals that we need to assume the non-trapping condition for δa only.

Using this, we prove that under the non-trapping assumption the linearization $\delta X_{a,f}$ is invertible in a stable way, if we know that it is injective. We provide conditions for injectivity: either $(\delta a, \delta f)$ are supported in a small enough set, or (a, f) satisfy the following analyticity conditions: Ba and u are analytic in $K \times S^1$. For the non-linear identification problem we prove local uniqueness near such a, f, in particular assuming the non-trapping condition for δa, and a conditional Hölder stability estimate.

As an example, we consider radial a and f. First, we consider $a = 0$ and f being equal to the characteristic function of the unit disk. The rays then are the concentric circles $|x| = R, 0 \leq R < 1$. K is non-trapping, if and only if no entire circle of that family lies in K. We also study the case of general radial a and f. There is no uniqueness in that case, and in fact, given any C_0^{∞} radial a, f, one can find a radial f_0, so that $X_a f = X_0 f_0$. This is a trapping case (without non-trapping support restrictions, except that all functions are supported in the unit disk) and demonstrates that in the trapping case we may lose the well-posedness of the problem.

Acknowledgements Partly supported by NSF.

Eigenvalues and Spectral Determinants on Compact Hyperbolic Surfaces

Alexander Strohmaier and Ville Uski

Compact hyperbolic surfaces are two dimensional oriented Riemannian manifolds of constant negative curvature -1. They can be realized as quotients $\Gamma \backslash \mathbb{H}$ of the upper half plane $\mathbb{H} = \{(x, y) \mid y > 0\}$ by a discrete hyperbolic co-compact subgroup $\Gamma \subset SL(2, \mathbb{R})$. By the uniformization theorem any two dimensional compact oriented Riemannian manifold of genus $g > 1$ admits exactly one metric of constant curvature -1 in its conformal class, so that in each genus $g > 2$ there exist many different hyperbolic metrics. The moduli space of hyperbolic metrics on a given 2-dimensional surface of genus $g > 1$ modulo diffeomorphisms can be described as the quotient of the Teichmüller space $\mathcal{T}_g$ by the mapping class group. Here the Teichmüller space is defined as the space of hyperbolic metrics on a given 2-dimensional surface of genus $g > 1$ modulo diffeomorphisms that are homotopic to the identity. The Teichmüller space is known to have dimension $6g - 6$ and we use an explicit description in terms of the so-called Fenchel Nielsen coordinates.

The description is based on the observation that for any three positive numbers ℓ_1, ℓ_2, ℓ_3 there is an up to isometry unique sphere with three holes (a so called Y-piece), equipped with a metric of curvature -1, such that the boundary components are geodesics and have lengths ℓ_1, ℓ_2, ℓ_3. Now any hyperbolic surface can be constructed by gluing $2g - 2$ such Y-pieces. This construction is unique once an order of glueing is fixed and once the twist for each boundary component is specified. Thus, given $3g-3$ length parameters and $3g-3$ angle parameters together with a discrete prescription on how the surface is to be glued from the Y-pieces fixes a concrete surface. The boundary geodesics will then be a system of simple closed geodesics on the surface X.

We give an algorithm that allows to compute the eigenvalues λ_i of the (positive) Laplace operator Δ on X with high accuracy and with rigorous error bounds. We

A. Strohmaier (✉) · V. Uski
Loughborough University, LE12 9BZ Leicestershire, UK
e-mail: a.strohmaier@lboro.ac.uk; V.Uski@lboro.ac.uk

D. Grieser et al. (eds.), *Microlocal Methods in Mathematical Physics and Global Analysis*, 105
Trends in Mathematics, DOI 10.1007/978-3-0348-0466-0_24, © Springer Basel 2013

also show how derived spectral quantities such as the spectral determinant and the Casimir energy can be computed from a finite part of the spectrum.

The starting point for our algorithm is the construction of basis functions $\Phi_j(\lambda)$ on each pair of pants Y that are discontinuous along one geodesic but that satisfy $\Delta\Phi_j(\lambda) = \lambda\Phi_j(\lambda)$. For the construction of the basis functions we use the fact that each Y-piece can be cut open along one geodesic connecting two boundary components and the resulting set embeds isometrically into a hyperbolic cylinder. The symmetry of the hyperbolic cylinder can then be used to solve the eigenvalue equation using separation of variables.

Gluing the surface X from the cut-open Y-pieces then leads to boundary conditions for the functions on the Y-pieces. If one can find a linear combination of basis functions that satisfy or are close to satisfying the boundary conditions, then λ is close to an eigenvalue. For a function $\Psi = \sum_j a_j \Phi_j$ we use the quadratic form $F_{-3/2,-1/2}(\Psi)^2$ which measures the difference of the function values and their normal derivatives along the curves using the Sobolev norms $H^{-3/2}$ and $H^{-1/2}$. Thus, if $F_{-3/2,-1/2}(\Psi) = 0$ the boundary conditions are satisfied exactly. We proved that there exists a constant $C > 0$ depending only on the geometry such that $\Delta\Psi = \lambda\Psi$ and $\delta = C\frac{F_{-3/2,-1/2}(\Psi)}{\|\Psi\|_{L^2(M)}} < 1$ imply that the interval $[\lambda - \epsilon, \lambda + \epsilon]$ contains an eigenvalue, where

$$\epsilon = \frac{(1+\lambda)\delta}{1-\delta}.$$

We managed to give an estimate for the constant C which is good enough to be used in numerical computations. Approximating the function Ψ by linear combinations of the basis functions, the task of minimizing $\frac{F_{-3/2,-1/2}(\Psi)}{\|\Psi\|_{L^2(M)}}$ for finite linear combinations of the Φ_j becomes a problem in linear algebra, namely finding the generalized singular value of two matrices.

The linear algebra problem can be set in such a way that the computation of the error in the above theorem remains rigorous. This is done by choosing the matrices in such a way that the generalized singular values are equal to the Simpson rule expression when computing the L^2-norms of differences of boundary data. Using known bounds on the fourth derivatives of eigenfunctions and the error of the Simpson approximation this gives rigorous estimates for the error made by discretization. It is important here that the basis functions Φ_j can be chosen orthonormal in the L^2-norm of a smaller hyperbolic cylinder fitted into the Y-piece.

In order to find eigenvalues we compute the smallest generalized singular value $\sigma(A(\lambda), B(\lambda))$ of a pair of matrices $A(\lambda)$ and $B(\lambda)$, where the dimensions of the matrices depend on a number N, reflecting the number of basis functions used, and on λ. For this special set of basis functions we were able to prove that

$$\sigma(A(\lambda), B(\lambda)) \leq \beta(N, \lambda') + C(\lambda)|\lambda - \lambda'|$$

if λ is close to an eigenvalue λ'. Since both constants $\beta(N, \lambda')$ and $C(\lambda)$ can be computed explicitly in a concrete geometric situation this allows to find all eigenvalues. The constant $\beta(N, \lambda')$ is decaying exponentially fast in N so that not many basis functions need to be used to obtain high accuracy.

We implemented this algorithm in Fortran for surfaces of arbitrary genus. As a test case we present here the Bolza surface, which is a surface of genus two that maximizes the order of the symmetry group. For this surface we find the first nonzero eigenvalue $\lambda = 3.8388872588421995185866622450435$ where we believe all decimal places to be correct. The mathematically rigorous error bound that also includes a crude error estimate of the Simpson rule gives an interval of size 0.00000001 around this point. We calculated the first 2,000 eigenvalues in order to compute the spectral determinant which we find to be roughly 4.722738 where again we believe all given decimal places to be correct. Changing parameters in Teichmüller space shows that both quantities are local maxima and are likely to be global maxima. Using our method to compute determinants is more accurate than the methods employed previously.

This method is closely related to spectral questions in microlocal analysis. Non-concentration along closed geodesics, boundedness of pseudodifferential operators, as well as analyticity of eigenfunctions on the Grauert tube, play a role in the effectiveness of the method. This short presentation is based on the article [1].

Reference

1. Alexander Strohmaier, Ville Uski, *Rigorous Computations of Eigenvalues and Spectral Zeta Functions and Zeta-Determinants on Hyperbolic Surfaces*, in preparation

Part IV
Wave Propagation and Topological Applications

A Support Theorem for a Nonlinear Radiation Field

Dean Baskin and António Sá Barreto

This note describes work in progress.

We consider the energy-critical semilinear wave equation on $\mathbb{R} \times \mathbb{R}^3$:

$$\Box u + |u|^4 u = 0, \tag{1}$$

$$(u, \partial_t u)|_{t=0} = (\phi, \psi).$$

This equation has been extensively studied (see, for example, the papers of Grillakis [4], Shatah-Struwe [5], Bahouri-Shatah [3], and Bahouri-Gérard [2]) and is known to possess unique solutions in

$$C^1\left(\mathbb{R}; L^2(\mathbb{R}^3)\right) \cap C^0\left(\mathbb{R}; \dot{H}^1(\mathbb{R}^3)\right) \cap L^5\left(\mathbb{R}; L^{10}(\mathbb{R})\right).$$

Bahouri and Gérard also showed that solutions to this equation exhibit *scattering*, i.e., given a solution u of equation (1), there are solutions $u_\pm$ of the linear problem $\Box u_\pm = 0$ such that

$$\|\nabla u(t) - \nabla u_\pm(t)\|_{L^2(\mathbb{R}^3)} + \|\partial_t u(t) - \partial_t u_\pm(t)\|_{L^2(\mathbb{R}^3)} \to 0 \text{ as } t \to \pm\infty.$$

If $(u_\pm, \partial_t u_\pm)_{t=0} = (\phi_\pm, \psi_\pm)$, we define the Moeller wave operator $\Omega_\pm$ by

$$\Omega_\pm(\phi_\pm, \psi_\pm) = (\phi, \psi).$$

D. Baskin (✉)
Northwestern University, Evanston, IL 60208, USA
e-mail: dbaskin@math.northwestern.edu

A. Sá Barreto
Purdue University, West Lafayette, IN 47907, USA
e-mail: sabarre@math.purdue.edu

D. Grieser et al. (eds.), *Microlocal Methods in Mathematical Physics and Global Analysis*, 111
Trends in Mathematics, DOI 10.1007/978-3-0348-0466-0_25, © Springer Basel 2013

We prove the following support theorem:

Theorem 1. *If $(\phi_\pm, \psi_\pm) \in C_c^\infty(\mathbb{R}^3)$ are radial and supported in $\{|z| \leq R\}$, then (ϕ, ψ) are also smooth, radial, and supported in $\{|z| \leq R\}$.*

To prove the theorem, we show that the Moeller wave operators can be defined in terms of Friedlander's wave operators, which are unitary translation representations of the semilinear wave group. In particular, we define the nonlinear forward and backward radiation fields for a solution u of equation (1) as rescaled restrictions of u to null infinity:

$$\mathcal{L}_\pm(\phi, \psi)(s, \theta) = \lim_{r \to \infty} \partial_s \left(r^{(n-1)/2} u(s \pm r, r\theta) \right),$$

which exist because $|u|^4 u \in L^1 L^2$.

The same formulae define the linear radiation fields. For a solution v of the linear wave equation $\Box v = 0$ with initial data $(\phi_\pm, \psi_\pm)$, the linear radiation field is given by

$$\mathcal{R}_\pm(\phi_\pm, \psi_\pm)(s, \theta) = \lim_{r \to \infty} \partial_s \left(r^{(n-1)/2} v(s \pm r, r\theta) \right).$$

The relationship between the nonlinear scattering operators and the Moeller wave operators is given by

$$\Omega_\pm = \mathcal{L}_\pm^{-1} \mathcal{R}_\pm.$$

Although Theorem 1 implies that $\Omega_\pm$ preserve supports for smooth compactly supported radial functions, their inverses generally do not. In particular, Theorem 1 is *not* a nonlinear Huygens' principle.

We prove the following support theorem for $\mathcal{L}_+$, which is equivalent in this setting to Theorem 1.

Theorem 2. *Suppose that $F \in C_c^\infty(\mathbb{R})$ vanishes for $|s| \geq R$ and satisfies $\int F = 0$. Then $F = \mathcal{L}_+(\phi, \psi)$ for radial $\phi, \psi \in C_c^\infty(\mathbb{R}^3)$, supported in $\{|z| \leq R\}$.*

The proof of this theorem relies on unique continuation results in $1 + 1$-dimensions and is unlikely to be true for non-radial data (see, for example, the paper of Alinhac-Baouendi [1]).

References

1. S. Alinhac and M. S. Baouendi. A nonuniqueness result for operators of principal type. *Math. Z.*, 220(4):561–568, 1995.
2. Hajer Bahouri and Patrick Gérard. High frequency approximation of solutions to critical nonlinear wave equations. *Amer. J. Math.*, 121(1):131–175, 1999.
3. Hajer Bahouri and Jalal Shatah. Decay estimates for the critical semilinear wave equation. *Annales de l'Institut Henri Poincaré - Analyse non linéaire*, 15(6):783–789, 1998.
4. Manoussos Grillakis. Regularity and asymptotic behaviour of the wave equation with a critical nonlinearity. *Ann. of Math. (2)*, 132(3):485–509, 1990.
5. Jalal Shatah and Michael Struwe. Well-posedness in the energy space for semilinear wave equations with critical growth. *Internat. Math. Res. Notices*, (7):303ff., approx. 7 pp. (electronic), 1994.

Propagation of Singularities Around a Lagrangian Submanifold of Radial Points

Nick Haber and András Vasy

This talk discusses the wavefront set of a solution u to $Pu = f$, where P is a pseudodifferential operator on a manifold with real-valued homogeneous principal symbol p, when the Hamilton vector field corresponding to p is radial on a Lagrangian submanifold contained in the characteristic set of P. According to a theorem of Duistermaat-Hörmander [2], singularities propagate along bicharacteristics of this Hamilton vector field. This theorem gives us no information about the wavefront set when the Hamilton vector field is radial.

Analysis takes place on X, an n-dimensional manifold without boundary. Let o be the 0-section of T^*X. Denote by $\mathcal{M} : T^*X \backslash o \times \mathbb{R}_+ \to T^*X \backslash o$ the natural dilation of the fibers of $T^*X \backslash o$: given $v \in T_x^* X, v \neq 0$, $\mathcal{M}((x, v), t) = (x, tv)$. We call a subset of $T^*X \backslash o$ conic if $\mathcal{M}$ acts on it.

Definition 1. We call the vector field $f(\cdot) \mapsto \frac{d}{dt}|_{t=0} f(\mathcal{M}(\cdot, t))$ the radial vector field. We say that H_p (the Hamilton vector field associated to symbol p) is *radial* at a point $q \in T^*X$ if H_p is a scalar multiple of the radial vector field at q, and we then call q a *radial point* of H_p.

If we choose local canonical coordinates (x, ξ) for T^*X, then H_p is radial at q if it is a scalar multiple of $\xi \cdot \partial_\xi$ at q. Equivalently, H_p is radial at q if dp is a scalar multiple of the canonical 1-form of T^*X, given in local canonical coordinates by $\sum_i \xi_i dx_i$.

Melrose [9] and Vasy [11] give a global analysis of the propagation of singularities around a Lagrangian submanifold of radial points. By adapting the standard positive commutator estimate proof of this theorem, we microlocalize these results. We let S^*X be the coshpere bundle at infinity, and $\kappa : T^*X \to S^*X$

N. Haber (✉) · A. Vasy
Department of Mathematics, Stanford University, Stanford, CA 94305, USA
e-mail: nhaber@math.stanford.edu; andras@math.stanford.edu

D. Grieser et al. (eds.), *Microlocal Methods in Mathematical Physics and Global Analysis*,   113
Trends in Mathematics, DOI 10.1007/978-3-0348-0466-0_26, © Springer Basel 2013

the quotient map. If H_p is radial at $\kappa^{-1}(q)$, then the images under κ of a set of flow lines of H_p have q as a limit point. Let Γ_q be the union of these images. In the following, we treat wavefront sets to be subsets of S^*X.

Theorem 1. *Given $P \in \Psi^m(X)$ with a real-valued homogeneous principal symbol p such that H_p is radial (and nonvanishing) on a conic Lagrangian submanifold $\Lambda \subset \dot{\Sigma}(P)$, then given $q \in \kappa(\Lambda)$, there exist $s_0, s_1 \in \mathbb{R}$ such that*

- *For $s < s_0$, if there is an open neighborhood $U_0 \subset S^*X$ of q disjoint from $\mathrm{WF}^{s-m+1}(Pu)$ and from $\Gamma_q \cap \mathrm{WF}^s(u)$, then $q \notin \mathrm{WF}^s(u)$.*
- *If $s > s_1$, then $q \notin \mathrm{WF}^{s_1}(u)$ implies $q \notin \mathrm{WF}^s(u) \backslash \mathrm{WF}^{s-m+1}(Pu)$.*

In particular, if $P - P^*$ has a homogeneous choice of principal symbol, then we can choose any $s_1 > s_0$. The values of s_0 and s_1 can be determined explicitly. Let ζ be a homogeneous elliptic symbol of order 1, defined locally in a neighborhood of $\kappa^{-1}(q)$. Let $\lambda = -H_p\zeta$. If we assume that $P - P^*$ has homogeneous principal symbol and take $\sigma_{m-1}(\frac{P-P^*}{2i})$ to be homogeneous, then

$$f(w) := \frac{\sigma_{m-1}(\frac{P-P^*}{2i})\zeta}{\lambda}(w)$$

is homogeneous of order 0, and thus a function on S^*X. It is then optimal to choose

$$s_0 = f(q) + \frac{m-1}{2}$$

and any $s_1 > s_0$. If we do not assume that $P - P^*$ has homogeneous principal symbol, then f is not homogeneous, and we must take a limit over neighborhoods about q. Explicitly,

$$s_0 := \sup_{U_0' \subset U_0 \text{ with } q \in U_0', \zeta_0 > 0} \left(\inf_{\{w \in U \mid \kappa(w) \in U_0', \zeta(w) > \zeta_0\}} f(w) + \frac{m-1}{2} \right)$$

and

$$s_1 > \inf_{U_0' \subset U_0, q \in U_0', \zeta_0 > 0} \left(\sup_{\{w \in U \mid \kappa(w) \in U_0', \zeta(w) > \zeta_0\}} f(w) + \frac{m-1}{2} \right).$$

These values can be derived as follows. We prove the above theorem with a positive commutator argument. Let b be the symbol of the commutant. This can be taken to have a simple form: $b = \chi\rho$, where χ is an order-0 cutoff, and ρ is a weight, depending on ζ and the regularity order s we wish to show. Around the Lagrangian submanifold, $\chi H_p\rho$ is the dominant term in $H_p b$. The sign of this term thus depends on whether the desired regularity is high or low. As the commutator must also absorb the subprincipal term $\frac{P-P^*}{2i}$, this term shifts the thresholds.

It should be emphasized that these results are completely local. That is, in order to conclude regularity for u at a point q in this Lagrangian submanifold, we need only have regularity for f in an arbitrarily small neighborhood of q. At times we also need regularity assumptions on u around Γ_q, and at other times we also need a priori lower regularity assumptions on u – it is important to note that these requirements are again local around q. Thus we do not, for instance, require regularity assumptions around the whole Lagrangian submanifold.

Under the nondegeneracy assumption $dp \neq 0$, the largest-dimensional subspace on which a Hamilton vector field can be radial is a Lagrangian submanifold. This occurs naturally in many applications, including geometric scattering theory. Indeed, this result generalizes a result in [9]. For the treatment of the opposite extreme, that is, that of an isolated radial point, see for instance [3–5].

Here is a simple example of such a situation. Working on $\mathbb{R}^n$, if we conjugate $\Delta - \lambda^2$ ($\lambda \neq 0$) by the Fourier transform, we get the multiplication operator $|x|^2 - \lambda^2$. Using canonical coordinates (x, ξ), this has Hamilton vector field $-2x \cdot \partial_\xi$, which is radial on the conormal bundle of $\{|x|^2 = \lambda^2\}$. This is a Lagrangian submanifold. Thus, regularity information for solutions to $|x|^2 - \lambda^2$ gives decay information for solutions to $\Delta - \lambda^2$. The threshold $s_0 = -\frac{1}{2}$ corresponds to, for $|x|^2 - \lambda^2$, the existence of delta functions, and for $\Delta - \lambda^2$, the existence of spherical waves. This analysis generalizes to the Laplacian on asymptotically euclidean spaces; see [9] for details.

A paper containing this result is in preparation and should appear shortly.

Acknowledgements Partially supported by the Department of Defense (DoD) through the National Defense Science and Engineering Graduate Fellowship (NDSEG) Program and a National Science Foundation Graduate Research Fellowship under Grant No. DGE-0645962. Partially supported by the NSF under Grant No. DMS-0801226 and from a Chambers Fellowship at Stanford University.

References

1. Jean-François Bony and Setsuro Fujiié and Ramond Thierry and Maher Zerzeri. Microlocal kernels of pseudodifferential operators at a hyperbolic fixed point. *J. Funct. Anal.*, 252(1): 68–125, 2007.
2. J. J. Duistermaat and L. Hörmander. Fourier Integral Operators, II. *Acta Math.* 128(3–4): 183–269, 1972.
3. Victor Guillemin and David Schaeffer. On a certain class of Fuchsian partial differential equations. *Duke Mathematical Journal*, 44(1):157–199, 1977.
4. Andrew Hassell, Richard Melrose, and András Vasy. Spectral and scattering theory for symbolic potentials of order zero. *Advances in Mathematics*, 181(1):1–87, 2004.
5. Andrew Hassell and Richard Melrose and András Vasy. Microlocal propagation near radial points and scattering for symbolic potentials of order zero. *Analysis and PDE*. 1:127–196, 2008.
6. Ira Herbst and Erik Skibsted. Quantum scattering for potentials independent of $|x|$: asymptotic completeness for high and low energies. *Communications in Partial Differential Equations*, 29(3–4):547 – 610, 2004.
7. Ira Herbst and Erik Skibsted. Absence of quantum states corresponding to unstable classical channels. *Annales Henri Poincaré*, 9(3):509–552, 2008.

8. Lars Hörmander. On the existence and regularity of solutions of linear pseudo-differential equations. *Enseignement Math. (2)*, 17: 99–163, 1971.
9. Richard Melrose. Spectral and Scattering theory for the Laplacian on asymptotically Euclidian spaces. *Spectral and scattering theory*, pages 85–130, 1994.
10. András Vasy. Asymptotic behavior of generalized eigenfunctions in N-body scattering. *J. Funct. Anal.*, 173(1):170–184, 1997.
11. András Vasy, with an appendix by Semyon Dyatlov. Microlocal analysis of asymptotically hyperbolic and Kerr-de Sitter spaces. arXiv:1012.4391v1 [math.AP].

Local Energy Decay for Several Evolution Equations on Asymptotically Euclidean Manifolds

Dietrich Häfner and Jean-François Bony

2000 Mathematics Subject Classification: 35L05, 35J10, 35P25, 58J45, 81U30.

This report is devoted to the study of the local energy decay for several evolution equations associated to long range metric perturbations of the Euclidean Laplacian on $\mathbb{R}^d$. We consider the following operator on $\mathbb{R}^d$, with $d \geq 1$,

$$P = -b\,\mathrm{div}(G\nabla b) = -\sum_{i,j=1}^{d} b(x)\frac{\partial}{\partial x_i}G_{i,j}(x)\frac{\partial}{\partial x_j}b(x),$$

where $b(x) \in C^\infty(\mathbb{R}^d)$ and $G(x) \in C^\infty(\mathbb{R}^d;\mathbb{R}^{d\times d})$ is a real symmetric $d \times d$ matrix. The C^∞ hypothesis is made mostly for convenience, much weaker regularity could actually be considered. We make an ellipticity assumption:

$$\exists C > 0,\ \forall x \in \mathbb{R}^d \qquad G(x) \geq CI_d \ \text{ and } \ b(x) \geq C, \tag{H1}$$

I_d being the identity matrix on $\mathbb{R}^d$. We also assume that P is a long range perturbation of the Euclidean Laplacian:

$$\exists \rho > 0,\ \forall \alpha \in \mathbb{N}^d \qquad |\partial_x^\alpha(G(x) - I_d)| + |\partial_x^\alpha(b(x) - 1)| \lesssim \langle x \rangle^{-\rho - |\alpha|}. \tag{H2}$$

D. Häfner (✉)
Institut Fourier-UMR5582 du CNRS, Université de Grenoble 1, 100 rue des Maths,
BP 74, 38402 St Martin d'Hères, France
e-mail: Dietrich.Hafner@ujf-grenoble.fr

J.-F. Bony
Institut de Mathématiques de Bordeaux, UMR 5251 du CNRS, Université Bordeaux 1,
351 cours de la Libération, 33405 Talence cedex, France
e-mail: Jean-Francois.Bony@math.u-bordeaux1.fr

D. Grieser et al. (eds.), *Microlocal Methods in Mathematical Physics and Global Analysis*, 117
Trends in Mathematics, DOI 10.1007/978-3-0348-0466-0_27, © Springer Basel 2013

For the high energy part, we will assume that

$$P \text{ is non-trapping.} \tag{H3}$$

In the following, $\| \cdot \|$ will always design the norm on $L^2(\mathbb{R}^d)$. We obtain local energy decay estimates for various evolution equations (see [2]).

Theorem 1. *Assume* (H1)–(H3) *and* $d \geq 2$. *For all* $\varepsilon > 0$, *we have*

(i) *for the wave equation*

$$\left\| \langle x \rangle^{1-d} \frac{\sin t \sqrt{P}}{\sqrt{P}} u \right\|_{H^1(\mathbb{R}^d)} \lesssim \langle t \rangle^{1-d+\varepsilon} \left\| \langle x \rangle^{d-1} u \right\|,$$

$$\left\| \langle x \rangle^{-d} (\partial_t, \sqrt{P}) \frac{\sin t \sqrt{P}}{\sqrt{P}} u \right\| \lesssim \langle t \rangle^{-d+\varepsilon} \left\| \langle x \rangle^d u \right\|.$$

(ii) *for the Klein–Gordon equation*

$$\left\| \langle x \rangle^{-d/2} e^{it\sqrt{1+P}} u \right\| \lesssim \langle t \rangle^{-d/2+\varepsilon} \left\| \langle x \rangle^{d/2} u \right\|.$$

(iii) *for the Schrödinger equation*

$$\left\| \langle x \rangle^{-d/2} e^{itP} u \right\| \lesssim |t|^{-d/2} \langle t \rangle^{\varepsilon} \left\| \langle x \rangle^{d/2} u \right\|_{H^{-d/2}(\mathbb{R}^d)}.$$

(iv) *for the capillary water wave equation*

$$\left\| \langle x \rangle^{-2d/3} e^{itP^{3/4}} u \right\| \lesssim |t|^{-2d/3} \langle t \rangle^{\varepsilon} \left\| \langle x \rangle^{2d/3} u \right\|_{H^{-d/3}(\mathbb{R}^d)}.$$

(v) *for the gravity water wave equation*

$$\left\| \langle x \rangle^{-2d} e^{itP^{1/4}} u \right\| \lesssim \langle t \rangle^{-2d+\varepsilon} \left\| \langle x \rangle^{2d} u \right\|_{H^d(\mathbb{R}^d)}.$$

Similar results are obtained for the fourth order Schrödinger equation in [2]. For compactly supported perturbations, the theory of resonances can be applied to obtain decay of the local energy (see [7]). In the potential case, one can use perturbation theory as in [6]. For small time dependent perturbations of the Minkowski metric outside obstacles, see [8]. In our setting, estimates of powers of the resolvent which imply local energy decay with other decay rates have been obtained in [4]. In the case of scattering manifolds, we refer to [9, 10].

In even dimensions, the estimates in i) are optimal modulo the loss of $\langle t \rangle^{\varepsilon}$. In odd dimensions and for short range perturbations, we have (see [3])

Theorem 2. *Assume* (H1)–(H3) *and* $d \geq 3$ *odd. Suppose furthermore* $\rho > \mu + 2$ ($\rho > \mu + 1$ *in dimension* $d = 3$) *with* $\mu \geq 0$. *For all* $\varepsilon > 0$, *we have*

$$\left\| \langle x \rangle^{-\mu-1-\varepsilon} \frac{\sin t \sqrt{P}}{\sqrt{P}} \langle x \rangle^{-\mu-1-\varepsilon} \right\| \lesssim \langle t \rangle^{-\mu}.$$

If $d = 3$ and $\mu \geq 1$, we can replace $\langle x \rangle^{-\mu-1-\varepsilon}$ by $\langle x \rangle^{-\mu-1/2-\varepsilon}$.

Our analysis separates into a low and high frequency analysis. For the low frequency part, we obtain the following general theorem without assuming the non-trapping condition (see [2]).

Theorem 3. *Assume* (H1)–(H2) *and* $d \geq 2$. *Let* f *be a real function such that* $f(x) = a_0 + a_1 x^\alpha + x^{\alpha+\nu} g(x)$ *with* $a_1 \neq 0$, $\alpha, \nu > 0$ *and* $g \in C^\infty(\mathbb{R})$. *Let* $\chi \in C_0^\infty(\mathbb{R})$ *be such that* $f'(x) > 0$ *for all* $x \in \mathrm{supp}\chi \cap]0, +\infty[$.

(i) *If* $0 < \alpha \leq 1$, *we have for all* $\varepsilon > 0$

$$\left\| \langle x \rangle^{-\frac{d}{2\alpha}} e^{itf(P)} \chi(P) \langle x \rangle^{-\frac{d}{2\alpha}} \right\| \lesssim \langle t \rangle^{-\frac{d}{2\alpha}+\varepsilon}.$$

(ii) *If* $\alpha > 1$, *we have for all* $\varepsilon > 0$

$$\left\| \langle x \rangle^{-\frac{d}{2}} e^{itf(P)} \chi(P) \langle x \rangle^{-\frac{d}{2}} \right\| \lesssim \langle t \rangle^{-\frac{d}{2\alpha}+\varepsilon}.$$

The proof of this theorem rests on Mourre theory (see [5]). More precisely, we obtain positive commutator estimates at the bottom of the spectrum and generalized Hardy type estimates. Our method also gives low frequency resolvent estimates (see [1]).

Theorem 4. *Assume* (H1)–(H2) *and* $d \geq 3$. *For all* $\alpha, \beta > 1/2$ *with* $\alpha + \beta > 2$, *we have*

$$\sup_{z \in \mathbb{C} \setminus \mathbb{R},\, |z| < 1} \left\| \langle x \rangle^{-\alpha} (P - z)^{-1} \langle x \rangle^{-\beta} \right\| \lesssim 1.$$

This estimate is false for the Euclidean Laplacian in dimension 3 if $\alpha \leq 1/2$ or $\beta \leq 1/2$ or $\alpha + \beta < 2$. Theorem 1 follows from Theorem 3 and the following general result at high frequency proved by semiclassical methods (see [2]).

Theorem 5. *Assume* (H1)–(H3) *and* $d \geq 1$. *Let* f *be a real function such that, for* $x \geq 1$, $f(x) = x^\alpha + x^{\alpha-\nu} g(x)$, *with* $\alpha, \nu > 0$ *and* $g(\frac{1}{x}) \in C^\infty([0, 1[)$.

(i) *For all* $\varphi \in C_0^\infty(]0, +\infty[)$ *and* $\mu \geq 0$, *we have*

$$\left\| \langle x \rangle^{-\mu} e^{itf(P)} \varphi(h^2 P) \langle x \rangle^{-\mu} \right\| \lesssim \langle t h^{1-2\alpha} \rangle^{-\mu},$$

uniformly for $h > 0$ *small enough and* $t \in \mathbb{R}$.

(ii) *For all* $\chi \in C_0^\infty(\mathbb{R})$ *equal to* 1 *on a large neighborhood of* 0 *and* $\mu \geq 0$,

$$\left\| \langle x \rangle^{-\mu} e^{itf(P)} (1-\chi)(P) u \right\|_{L^2(\mathbb{R}^d)} \lesssim \begin{cases} \langle t \rangle^{-\mu} \left\| \langle x \rangle^{\mu} u \right\|_{H^{\mu-2\alpha\mu}(\mathbb{R}^d)} & \text{for } \alpha \leq 1/2, \\[2ex] |t|^{-\mu} \left\| \langle x \rangle^{\mu} u \right\|_{H^{\mu-2\alpha\mu}(\mathbb{R}^d)} & \text{for } \alpha > 1/2. \end{cases}$$

References

1. J.-F. Bony and D. Häfner, *Low frequency resolvent estimates for long range perturbations of the Euclidean Laplacian*, Math. Res. Lett. **17** (2010), no. 2, 301–306.
2. J.-F. Bony and D. Häfner, *Local energy decay for several evolution equations on asymptotically Euclidean manifolds*, preprint arXiv:1008.2357.
3. J.-F. Bony and D. Häfner, *Improved local energy decay for the wave equation on asymptotically Euclidean odd dimensional manifolds*, in preparation.
4. J.-M. Bouclet, *Low frequency estimates and local energy decay for asymptotically Euclidean Laplacians*, preprint arXiv:1003.6016.
5. W. Hunziker, I. M. Sigal, and A. Soffer, *Minimal escape velocities*, Comm. Partial Differential Equations **24** (1999), no. 11–12, 2279–2295.
6. A. Jensen and T. Kato, *Spectral properties of Schrödinger operators and time-decay of the wave functions*, Duke Math. J. **46** (1979), no. 3, 583–611.
7. P. Lax and R. Phillips, *Scattering theory*, second ed., Pure and Applied Mathematics, vol. 26, Academic Press Inc., 1989, With appendices by C. Morawetz and G. Schmidt.
8. J. Metcalfe and D. Tataru, *Decay estimates for variable coefficient wave equations in exterior domains*. Advances in phase space analysis of partial differential equations, 201–216, Progr. Nonlinear Differential Equations Appl., 78, Birkhäuser Boston, 2009.
9. A. Vasy and J. Wunsch, *Positive commutators at the bottom of the spectrum*, J. Funct. Anal. 259 (2010), no. 2, 503–523.
10. I. Rodnianski and T. Tao, *Effective limiting absorption principle, and applications*, preprint arXiv:1105.0873.

Rayleigh Surface Waves and Geometric Pseudo-differential Calculus

Sönke Hansen

1991 Mathematics Subject Classification: 35Q74, 35S05.

Applying microlocal analysis we study Rayleigh-type elastic surface waves. For precise statements, proofs, and additional references see [3].

By Hooke's law, the stress (=force) tensor of an elastic material is proportional to strain, $s = Ce$. The strain e corresponding to an (infinitesimal) displacement field u is the deformation of the metric tensor g of the 3-dimensional Riemannian manifold M filled by the medium, $e = \operatorname{Def} u = \mathcal{L}_u g / 2$. Stress and strain are symmetric 2-tensors. The elasticity tensor C is a 4-tensor defining a scalar product on symmetric 2-tensors. It vanishes on antisymmetric tensors. The elastic properties of the material are encoded in C. The elasticity operator L and the traction Tu on the boundary $X = \partial M$ are defined by

$$\int_M (\operatorname{Def} u \mid \operatorname{Def} v)_C \, \mathrm{d}V_M = \int_M (Lu \mid v) \, \mathrm{d}V_M + \int_X (Tu \mid v) \, \mathrm{d}V_X,$$

Thus $L = \operatorname{Def}^* \operatorname{Def} = -\nabla^* C \nabla$.

We are interested in solutions with vanishing traction, $Tu = 0$ on X, of the wave equation, $Lu - D_t^2 u = 0$, and of the spectral problem, $h^2 Lu - u = 0$ as $h \downarrow 0$. The semiclassical principal symbol of $h^2 L - \operatorname{Id}$ equals $c(\xi) - \operatorname{Id} \in \operatorname{End}(T_x M)$ at $\xi \in T_x^* M$. The acoustic tensor $c(\xi) = c(\xi, \xi)$ is a contraction of the elasticity tensor, $c(\xi, \eta) = \xi.C.\eta$. If the medium is isotropic, i.e., if C is $SO(3)$-invariant, then the elastodynamic system in the interior is a system of real principal type in the sense of Dencker [2]. It follows from this that Lagrangian distribution solutions can be constructed by solving invariantly defined transport equations.

S. Hansen (✉)
Institut für Mathematik, Universität Paderborn, 33095 Paderborn, Germany
e-mail: soenke@math.upb.de

D. Grieser et al. (eds.), *Microlocal Methods in Mathematical Physics and Global Analysis*,   121
Trends in Mathematics, DOI 10.1007/978-3-0348-0466-0_28, © Springer Basel 2013

The elliptic boundary region $\mathcal{E} \subset T^*X$ consists of all $\xi \in T^*X = \nu^\perp \subset T^*M$ such that $\ell(s) := c(\xi + s\nu) - \mathrm{Id}$ is positive definite for real s, ν the unit exterior conormal field. Rays from the interior do not intersect $\mathcal{E}$. From the theory of self-adjoint matrix polynomials it follows that there is a unique factorization $\ell(s) = (s - q^*)c(\nu)(s - q)$ such that the spectrum of q is contained in the lower half-plane. Moreover,

$$q \oint \ell(s)^{-1} \mathrm{d}s = \oint s\, \ell(s)^{-1} \mathrm{d}s,$$

where the contour of integration encloses only that part of the spectrum of ℓ which lies in the lower half-plane.

The above factorization is used to construct, microlocally over $\mathcal{E}$, a Dirichlet parametrix B_h and the DN map

$$Z_h \equiv T B_h : u|_X \mapsto T u|_X, \quad h^2 L u - u = 0.$$

The principal symbol z of Z_h is called the surface impedance tensor in applied physics. It is self-adjoint, and $z(\xi)$ is positive definite for sufficiently large $|\xi|$. Rayleigh waves arise from the characteristic set $\Sigma := \{\det z(\xi) = 0\} \subset \mathcal{E}$. In the applied physics literature, an important advance in the theory of surface waves of general elastic media was made by Barnett and Lothe [4]. Their proof of what is called the uniqeness of surface waves shows that, when translated to microlocal analysis, Z_h is a real principal type system. We sketch the argument. A calculation of the principal symbol of the composition $T B_h$ gives the following formula for z:

$$z \oint \ell(s)^{-1} \mathrm{d}s = i \oint c(\nu, s\nu + \xi)\, \ell(s)^{-1} \mathrm{d}s.$$

Using the residue calculus one derives

$$\mathrm{Re}\, z \int_{-\infty}^{\infty} \ell(s)^{-1} \mathrm{d}s = \pi\, \mathrm{Id},$$

hence the positive definiteness of $\mathrm{Re}\, z$ everywhere in $\mathcal{E}$. If z had only one positive eigenvalue with eigenvector ν, then $\mathrm{Re}\, z$ would be negative semi-definite on real vectors orthogonal to the real and to the imaginary part of ν. Since z is 3×3, it follows that z has at most one non-positive eigenvalue, implying the real principal type property.

In the isotropic case, Taylor [6] showed how in microlocal analysis the Rayleigh wave phenomenon is understood as propagation of singularities for the DN map over $\mathcal{E}$. Here $\Sigma = \{c_r|\xi| = 1\} \neq \emptyset$, where $c_r > 0$ is the Rayleigh wave speed which is strictly smaller than the interior wave speeds. Cardoso and Popov [1] and Stefanov [5] constructed Rayleigh quasimodes for isotropic media.

We generalize the construction of Rayleigh quasimodes to general elastic media, not nesessarily isotropic. To guarantee existence, we assume that Σ intersects each halfray $\mathbb{R}_+\xi$, and also that the line bundle $\ker z \to \Sigma$ is trivial. The first assumption implies $\Sigma = \{p = 1\}$ with $0 < p$ smooth and positively homogeneous of degree one. Following Stefanov's approach we show that, microlocally near Σ, the DN map intertwines into a scalar eigenvalue problem; essentially,

$$Z_h J_h \equiv J_h (P - h^{-1}).$$

Here P is a self-adjoint first order scalar pseudo-differential operator on X with principal symbol p. In particular, P is elliptic and bounded from below. The eigenfunctions of P are mapped by the intertwiner J_h and the Dirichlet parametrix B_h into a sequence of quasimodes. This leads to (sub-)sequences of eigenvalues and resonances for zero traction boundary and scattering problems, respectively.

The asymptotic behaviour of the eigenvalues of the operator P is affected by the subprincipal symbol. We derive an explicit formula for the subprincipal symbol p_s of P which is new even for isotropic elasticity. The curvature of the boundary X enters into the formula. Also p_s depends on the choice of a global section of $\ker z \to \Sigma$ because the construction of the intertwiner J_h does. When changing to a different section, a Poisson bracket $\{p, \varphi\}$ is added to the formula for p_s. Such terms are seen to cancel in asymptotic formulas for eigenvalues, however.

For computing the subprincipal symbol of P we use the geometric pseudo-differential calculus developed by Widom [7] and others. In this calculus, differential geometric structure enters into the quantization rule. Symbols are quantized as follows:

$$\mathrm{Op}(a)u(x) = \int_{T_x^* X} \int_{T_x X} e^{-i\langle \eta, v \rangle} a(x, \eta) \tau_{[x \leftarrow \exp_x v]} u(\exp_x v) \mathrm{d}v \mathrm{d}\eta.$$

Here exp denotes the exponential map of the Levi-Civita connection, and $\tau_{[x \leftarrow y]}$ is the parallel transport map from y to x along the shortest geodesic. The full symbol of the composition $\mathrm{Op}_h(a \sharp b) = \mathrm{Op}_h(a)\,\mathrm{Op}_h(b)$ is given by

$$a \sharp b \sim ab - ih\,\mathrm{tr}\,\nabla^v a\,\nabla^h b + \dots,$$

where the horizontal derivative ∇^h is defined using the connection. This allows to conveniently track symbols of systems of pseudo-differential operators such as Z_h down to subprincipal level.

References

1. Fernando Cardoso and Georgi Popov, *Rayleigh quasimodes in linear elasticity*, Comm. Partial Differential Equations **17** (1992), no. 7–8, 1327–1367.
2. Nils Dencker, *On the propagation of polarization sets for systems of real principal type*, J. Funct. Anal. **46** (1982), 351–372.
3. Sönke Hansen, *Rayleigh-type surface quasimodes in general linear elasticity*, 2010, 46 pages, arXiv:1008.2930, to appear in Analysis&PDE.

4. J. Lothe and D. M. Barnett, *On the existence of surface-wave solutions for anisotropic half-spaces with free surface*, J. Applied Physics **47** (1976), 428–433.
5. Plamen Stefanov, *Lower bounds of the number of the Rayleigh resonances for arbitrary body*, Indiana Univ. Math. J. **49** (2000), no. 1, 405–426.
6. M. E. Taylor, *Rayleigh waves in linear elasticity as a propagation of singularities phenomenon*, Partial Differential Equations and Geometry (C.I. Byrnes, ed.), Marcel Dekker, 1979, pp. 273–291.
7. Harold Widom, *A complete symbolic calculus for pseudodifferential operators*, Bull. Sci. Math. (2) **104** (1980), no. 1, 19–63.

Topological Implications of Global Hypoellipticity

Gerardo A. Mendoza

I collect here some results concerning implications of a topological nature of conditions such as ellipticity or hypoellipticity of a differential or pseudodifferential operator. Results of this kind help to better understand the scope of hypotheses of such essentially analytic conditions. This is of interest, in particular, in the case of complexes of differential operators, whether elliptic or not (for example, CR complexes). Such complexes have been the subject of extensive investigation by many authors, one of the most remarkable results in the theory being that of Cordaro and Hounie [1] on local solvability (the validity of the Poincaré Lemma) for a certain non-elliptic complex.

The following theorem is part of joint work in progress (research partially supported by FAPESP, contract nr. 2008/56767-0) with A.P. Bergamasco and S.L. Zani [2, 3] on topological restrictions imposed by the assumption of global C^∞-hypoellipticity.

Theorem 1. *Suppose $\mathcal{M}$ is a closed orientable connected surface, $E, F \to \mathcal{M}$ are line bundles, and*

$$P : C^\infty(\mathcal{M}; E) \to C^\infty(\mathcal{M}; F) \tag{1}$$

is a first order differential operator of principal type. If P is globally hypoelliptic, then

$$c(F) - c(E) = \pm\mathbf{e}(\mathcal{M}).$$

Here $c(E)$ is the total Chern class of E and $\mathbf{e}(\mathcal{M})$ is the Euler class of $\mathcal{M}$. Global hypoellipticity of course means that P has the property that if $u \in C^{-\infty}(\mathcal{M}; E)$ and $Pu \in C^\infty(\mathcal{M}; F)$ then in fact $u \in C^\infty(\mathcal{M}; E)$. Finally, principal type is meant here in the classical sense: the restriction of the principal symbol of P to any fiber $T_x^*\mathcal{M}$ of $T^*\mathcal{M}$ is nonzero as a (linear) function $T_x\mathcal{M} \to \mathrm{Hom}(E_x, F_x)$. The proof of Theorem 1 uses a microlocal argument concerning solvability of the transpose of P

G.A. Mendoza (✉)
Department of Mathematics, Temple University, Philadelphia, PA 19122, USA
e-mail: gmendoza@temple.edu

D. Grieser et al. (eds.), *Microlocal Methods in Mathematical Physics and Global Analysis*,   125
Trends in Mathematics, DOI 10.1007/978-3-0348-0466-0_29, © Springer Basel 2013

which ends up allowing us to deform P to an elliptic differential operator of order 1 using an idea from [6]. Once this is accomplished, the result is a consequence of Theorem 4 discussed below.

Theorem 1 generalizes the following theorem of Hounie [4]:

Theorem 2. *Suppose $\mathcal{M}$ is a closed orientable smooth surface and L is a vector field on $\mathcal{M}$ of principal type. If L is, as a differential operator, globally hypoelliptic, then $\mathcal{M}$ is a torus.*

The assumption that L is of principal type is equivalent to the statement that L is nowhere zero. Thus if L is a real vector field, then this implies immediately that $\mathcal{M}$ is a torus. However L may be a complex vector field, so Hounie's theorem is not immediate since the complexification of the tangent bundle of any manifold admits a globally defined nowhere zero vector field; a hypothesis such as global hypoellipticity is needed. Theorem 1 reduces to Hounie's theorem when E and F are the trivial vector bundle and P is has no zeroth order term (invariantly, P annihilates the constants).

Next is a theorem in which one reaches the same conclusion as in Theorem 1, starting with a different hypothesis. First some concepts. Let $\mathcal{M}$ be an arbitrary smooth paracompact manifold and $\iota : \mathcal{V} \hookrightarrow \mathbb{C}T\mathcal{M}$ a subbundle. Then there is an associated differential operator $\mathbb{D} : C^\infty(\mathcal{M}) \to C^\infty(\mathcal{M}; \mathcal{V}^*)$, namely if f is a smooth function, let $\mathbb{D}f = \iota^* df$. In other words, $\mathbb{D}f$ is the restriction of df to $\mathcal{V}$. When $\mathcal{V}$ is involutive, $\mathbb{D}$ is the first operator in a complex of first order differential operators on the exterior powers of $\mathcal{V}^*$ (see Treves [7]). Many natural differential complexes arise in this manner. Let $m = \dim \mathcal{M} - \mathrm{rk}\, \mathcal{V}$. The subbundle $\mathcal{V}$ is said to be a hypo-complex structure (see Treves, op cit.) if for each $x \in \mathcal{M}$ there exists some open neighborhood $\mathcal{U}$ of x and a C^∞ function $Z : U \to \mathbb{C}^m$ whose components Z_i satisfy $\mathbb{D}Z_i = 0$ over $\mathcal{U}$ and have independent differentials at x, with the property that for any u such that $\mathbb{D}u = 0$ near x there is h holomorphic near $Z(x)$ such that $u = h \circ Z$ near x. The following result [6, Theorem 7.3] was part of a general analysis of subbundles of $\mathbb{C}T\mathcal{M}$ carried out in joint work with H. Jacobowitz also aimed at getting a better sense of the analytical conditions that can be placed on complexes of the kind just described.

Theorem 3. *Suppose $\mathcal{M}$ is an orientable two-manifold and $\mathcal{V} \subset \mathbb{C}T\mathcal{M}$ is a hypo-complex subbundle (with $\mathrm{rk}\, \mathcal{V} = 1$). Then there exists a smooth family of subbundles $\mathcal{V}_t \subset \mathbb{C}T\mathcal{M}$, $0 \le t \le 1$ in which $\mathcal{V}_0 = \mathcal{V}$ and $\mathcal{V}_t$ is a holomorphic structure on $\mathcal{M}$ for each $t > 0$.*

This theorem can be viewed as intermediate between Theorems 1 and 2. Indeed, in the terminology of the first theorem, we have again that E is the trivial line bundle as in Theorem 2 but now $F = \mathcal{V}$. The conclusion of Theorem 1, namely that $\mathbf{c}_1(\mathcal{V}) = \pm \mathbf{e}(\mathcal{M})$, holds here because the Euler class of $\mathcal{M}$ is, except for sign (corresponding to choice a of orientation), the first Chern class of a holomorphic or antiholomorphic structure on $\mathcal{M}$.

With the stronger assumption of ellipticity we have the following correspondingly stronger result obtained in collaboration with Jacobowitz [5]

Theorem 4. *Let $\mathcal{M}$ be a connected compact manifold and $E, F \to \mathcal{M}$ complex vector bundles. If there is an elliptic classical pseudodifferential operator P : $C^\infty(\mathcal{M}; E) \to C^\infty(\mathcal{M}; F)$ then $c(F) - c(E) = k\mathbf{e}(\mathcal{M})$ for some $k \in \mathbb{Z}$.*

This is shown using the Gysin sequence associated with the cosphere bundle of $\mathcal{M}$, see [5]. A somewhat weaker result can be obtained if $\mathcal{M}$ is orientable using a Mayer-Vietoris sequence, as follows. Let p be the principal symbol of an elliptic pseudodifferential operator as in the statement of the theorem. Thus $p : \dot{\pi}^* E \to \dot{\pi}^* F$ is an isomorphism (where $\dot{\pi} : T^*\mathcal{M} \setminus 0 \to \mathcal{M}$ is the projection). Suppose for a moment that $\mathcal{M}$ admits a global nonvanishing (continuous) differential one-form α. The image of $\mathcal{M}$ by α is of course diffeomorphic to $\mathcal{M}$ and the isomorphism $p|_{\alpha(\mathcal{M})} : \dot{\pi}^* E|_{\alpha(\mathcal{M})} \to \dot{\pi}^* F|_{\alpha(\mathcal{M})}$ therefore descends to an isomorphism $E \to F$. That is, under the hypotheses of the theorem, if the Euler class of $\mathcal{M}$ vanishes, then $E = F$. Now, for general $\mathcal{M}$, pick $x \in \mathcal{M}$ arbitrarily, and let $\mathcal{U}$ be a neighborhood of x diffeomorphic to a ball. Then E is isomorphic to F over $\mathcal{M} \setminus \{x\}$ as well as over $\mathcal{U}$, since both these manifolds have vanishing Euler characteristic. The maps

$$H^{2q}(\mathcal{M}) \to H^{2q}(\mathcal{M} \setminus \{x\}) \oplus H^{2q}(\mathcal{U})$$

in the Mayer-Vietoris sequence in integral cohomology for the pair $\mathcal{M} \setminus \{x\}, \mathcal{U}$ are injective when $0 < 2q < \dim \mathcal{M}$, so since the image of $c_q(F) - c_q(E)$ in $H^{2q}(\mathcal{M} \setminus \{x\}) \oplus H^{2q}(\mathcal{U})$ vanishes, $c_q(E) - c_q(F) = 0$ when $2q < \dim \mathcal{M}$ (here $c_q(E)$ is the q-th Chern class of E). So $c(F) - c(E)$ is either 0 (when $\dim \mathcal{M}$ is odd) or a homogeneous class of top degree (when $\dim \mathcal{M}$ is even). The theorem asserts that the latter is proportional to the Euler class of $\mathcal{M}$.

Suppose now that $\mathcal{M}$ is a closed connected orientable surface, let $\mathbf{e}$ be the Euler class of $\mathcal{M}$ and let E and F be line bundles over $\mathcal{M}$. If the operator (1) is an elliptic differential differential operator, then $c_1(F) - c_1(E) = k\mathbf{e}$, $k \in \mathbb{Z}$. The number $|k|$ is the order of P (see [5, Corollary 2.5]). Thus Theorem 1 includes Theorem 4 in the restricted context (in dimension and rank) of the former.

References

1. P. Cordaro, J. Hounie, *Local solvability for a class of differential complexes*, Acta Math. **187** (2001), no. 2, 191–212.
2. A. P. Bergamasco, S. L. Zani, *Global analytic regularity for structures of co-rank one*, Comm. Partial Differential Equations **33** (2008), no. 4–6, 933–941.
3. _____, Globally analytic hypoelliptic vector fields on compact surfaces. Proc. Amer. Math. Soc. **136** (2008), no. 4, 13051310.
4. J. Hounie, *Globally hypoelliptic vector fields on compact surfaces*, Communications in Partial Differential Equations, **7** (1982), no. 4, 343–370.
5. H. Jacobowitz and G. A. Mendoza, *Elliptic equivalence of vector bundles*. Indiana Univ. Math. J. **51** (2002), 705–725.
6. _____, *Subbundles of the complexified tangent bundle*, Trans. Amer. Math. Soc. **355** (2003), 4201–4222.
7. F. Treves, *Hypo-analytic structures. Local theory*, Princeton Mathematical Series, **40**, Princeton University Press, Princeton, NJ, 1992.

Chern-Simons Line Bundle on Teichmüller Space

Sergiu Moroianu and Colin Guillarmou

We report in this note on our recent results about Chern-Simons invariants of certain geometrically finite hyperbolic 3-manifolds X. The ends of X can be either funnels or rank 2 cusps, but we focus here on *convex co-compact* hyperbolic manifolds, which are conformally compactifiable to a smooth manifold with boundary. More precisely, let M be a Riemann surface of genus ≥ 2 with a hyperbolic metric h_0, and A an endomorphism of TM satisfying $\mathrm{div}_{h_0} A = 0$, and $\mathrm{Tr}(A) = -\frac{1}{2}\mathrm{scal}_{h_0}$. A *hyperbolic funnel* is some collar $(0, \epsilon)_x \times M$ equipped with a metric

$$g = \frac{dx^2 + h(x)}{x^2}, \tag{1}$$

where $h(x) \in C^\infty(M, S_+^2 T^* M)$, and $h(x) = h_0\left((\mathrm{Id} + \frac{x^2}{2}A)\cdot, (\mathrm{Id} + \frac{x^2}{2}A)\cdot\right)$. The metric g on the funnel is of constant sectional curvature -1, and every end of X is isometric to such a hyperbolic funnel. The funnels have a conformal boundary, which is a disjoint union of compact Riemann surfaces forming the *conformal boundary* M of X. The deformation space of X is essentially the deformation space of its conformal boundary, i.e. Teichmüller space $\mathcal{T}$. A couple (h_0, A_0) can be considered as an element of $T_{h_0}^* \mathcal{T}$, if $A_0 = A - \frac{1}{2}\mathrm{tr}(A)\mathrm{Id}$ is the trace-free part of the divergence-free tensor A. We therefore identify the tangent bundle $T^*\mathcal{T}$ of $\mathcal{T}$ with the set of hyperbolic funnels modulo the action of the diffeomorphism group $\mathcal{D}_0(M)$, acting trivially in the x variable.

S. Moroianu (✉)
Institutul de Matematică al Academiei Române, P.O. Box 1-764, RO-014700 Bucharest, Romania
e-mail: moroianu@alum.mit.edu

C. Guillarmou
DMA, U.M.R. 8553 CNRS, Ecole Normale Supérieure, 45 rue d'Ulm,
F 75230 Paris cedex 05, France
e-mail: cguillar@dma.ens.fr

D. Grieser et al. (eds.), *Microlocal Methods in Mathematical Physics and Global Analysis*, 129
Trends in Mathematics, DOI 10.1007/978-3-0348-0466-0_30, © Springer Basel 2013

The renormalized volume of (X, g) is defined by

$$\mathrm{Vol}_R(X) := \mathrm{FP}_{\epsilon \to 0} \int_{x > \epsilon} \mathrm{vol}_g$$

where FP means finite-part (i.e. the coefficient of ϵ^0 in the asymptotic expansion as $\epsilon \to 0$). Here x is the distinguished boundary-defining function appearing in (1).

Let ω be the so(3)-valued Levi-Civita connection 1-form on X in an oriented orthonormal frame $S = (S_1, S_2, S_3)$. Let $\theta := \omega + iT$; here T is the so(3)-valued 1-form defined by $T_{ij}(V) := g(V \times S_j, S_i)$ and $\times$ is the vector product with respect to the metric g. We define

$$\mathrm{CS}(g, S) := -\tfrac{1}{16\pi^2} \mathrm{FP}_{\epsilon \to 0} \int_{x > \epsilon} \mathrm{Tr}(\omega \wedge d\omega + \tfrac{2}{3}\omega \wedge \omega \wedge \omega); \qquad (2)$$

$$\mathrm{CS}^{\mathrm{PSL}_2(\mathbb{C})}(g, S) := -\tfrac{1}{16\pi^2} \mathrm{FP}_{\epsilon \to 0} \int_X \mathrm{Tr}(\theta \wedge d\theta + \tfrac{2}{3}\theta \wedge \theta \wedge \theta). \qquad (3)$$

We ask that S be even to the first order at $\{x = 0\}$. Equipped with the conformal metric $\hat{g} := x^2 g$, the manifold X extends to a smooth Riemannian manifold $\overline{X} = X \cup M$ with boundary M. The Chern-Simons invariant $\mathrm{CS}(\hat{g}, \hat{S})$ is therefore well defined if $\hat{S} = x^{-1}S$ is an orthonormal frame for $\hat{g}$.

Proposition 1. *On a convex co-compact hyperbolic 3-manifold (X, g) one has $\mathrm{CS}(g, S) = \mathrm{CS}(\hat{g}, \hat{S})$, and*

$$\mathrm{CS}^{\mathrm{PSL}_2(\mathbb{C})}(g, S) = -\tfrac{i}{2\pi^2}\mathrm{Vol}_R(X) + \tfrac{i}{4\pi}\chi(M) + \mathrm{CS}(g, S) \qquad (4)$$

where $\chi(M)$ is the Euler characteristic of the conformal boundary M.

There exists a smooth map Φ from $\mathcal{T}$ to the set of geometrically finite hyperbolic metrics on X (up to diffeomorphisms of X) such that the conformal boundary of $\Phi(h)$ is (M, h) for any $h \in \mathcal{T}$. The subgroup Mod_X of the mapping class group Mod consisting of elements which extend to diffeomorphisms on $\overline{X}$ homotopic to the identity acts freely, properly discontinuously on $\mathcal{T}$ and the quotient is a complex manifold of dimension $3|\mathbf{g}| - 3$. The map Φ is invariant under the action of Mod_X and the deformation space $\mathcal{T}_X$ of X is identified with a quotient of the Teichmüller space, $\mathcal{T}_X = \mathcal{T}/\mathrm{Mod}_X$.

Since X is not closed, $e^{2\pi i \mathrm{CS}(g,S)}$ depends on the choice of the frame S near the conformal boundary, so $e^{2\pi i \mathrm{CS}^{\mathrm{PSL}_2(\mathbb{C})}}$ and $e^{2\pi i \mathrm{CS}}$ are not numerical invariants but rather sections in a complex line bundle $\mathcal{L}$ over the Teichmüller space $\mathcal{T}$.

Theorem 1. *There exists a holomorphic Hermitian line bundle $\mathcal{L}$ over $\mathcal{T}$ equipped with a Hermitian connection $\nabla^{\mathcal{L}}$, with curvature given by $\frac{i}{8\pi}$ times the Weil-Petersson symplectic form ω_{WP} on $\mathcal{T}$. The bundle $\mathcal{L}$ with its connection descend* to $\mathcal{T}_X$ and if $g_h = \Phi(h)$ is the geometrically finite hyperbolic metric with conformal boundary $h \in \mathcal{T}$, then $h \to e^{2\pi i \mathrm{CS}(g_h, \cdot)}$ is a global section of $\mathcal{L}$.

Since funnels can be identified to elements in $T^*\mathcal{T}$, the map Φ described above induces a section σ of the bundle $T^*\mathcal{T}$ (which descends to $T^*\mathcal{T}_X$) by assigning to $h \in \mathcal{T}$ the funnels of $\Phi(h)$. The image of σ

$$\mathcal{H} := \{\sigma(h) \in T^*\mathcal{T}_X, h \in \mathcal{T}_X\}$$

identifies the set of geometrically finite hyperbolic metrics on X as a graph in $T^*\mathcal{T}_X$. Define a modified connection

$$\nabla^\mu := \nabla^{\mathcal{L}} + \tfrac{2}{\pi}\mu^{1,0} \tag{5}$$

on the pull-back of $\mathcal{L}$ to $T^*\mathcal{T}$, where $\mu^{1,0}$ is the $(1,0)$ part of the Liouville 1-form μ on $T^*\mathcal{T}$. The connection descends to $T^*\mathcal{T}_X$, and it is not Hermitian, but ∇^μ and $\nabla^{\mathcal{L}}$ induce the same holomorphic structure on $\mathcal{L}$.

Theorem 2. *For $V \in T(T^*\mathcal{T})$ tangent to $\mathcal{H}$, we have $\nabla^\mu_V e^{2\pi i \mathrm{CS}^{\mathrm{PSL}_2(\mathbb{C})}} = 0$.*

The curvature of ∇^μ vanishes on $\mathcal{H}$ by Theorem 2 while the curvature of $\nabla^{\mathcal{L}}$ is $\frac{i}{8\pi}\omega_{\mathrm{WP}}$ (by Theorem 1). By considering the real and imaginary parts of these curvature identities, we obtain

Corollary 1. *The manifold $\mathcal{H}$ is Lagrangian in $T^*\mathcal{T}_X$ for the Liouville symplectic form μ and $d(\mathrm{Vol}_R \circ \sigma) = -\frac{1}{4}\mu$ on $\mathcal{H}$. The renormalized volume is a Kähler potential for the Weil-Petersson metric on $\mathcal{T}_X$:*

$$\bar\partial\partial(\mathrm{Vol}_R \circ \sigma) = \tfrac{i}{16}\omega_{\mathrm{WP}}.$$

Finally we relate the Chern-Simons line bundle $\mathcal{L}$ to the Quillen determinant line bundle $\det \partial = \Lambda^{\mathbf{g}}(\mathrm{coker}\,\partial)$ of ∂ on functions in the particular case of Schottky hyperbolic manifolds of genus g. Once we choose a marking of M by disjoint simple closed curves $\alpha_1, \ldots, \alpha_{\mathbf{g}}$ we obtain a canonical section $\varphi := \varphi_1 \wedge \cdots \wedge \varphi_{\mathbf{g}}$ where φ_j are holomorphic 1-forms on M normalized through the requirement $\int_{\alpha_j} \varphi_k = \delta_{jk}$.

Theorem 3. *Over the Schottky space, there is an explicit isometric isomorphism of holomorphic Hermitian line bundles between the inverse $\mathcal{L}^{-1}$ of the Chern-Simons line bundle and the 6-th power $(\det \partial)^{\otimes 6}$ of the determinant line bundle $\det \partial$, given by*

$$(F\varphi)^{\otimes 6} \mapsto e^{-2\pi i \mathrm{CS}^{\mathrm{PSL}_2(\mathbb{C})}}.$$

Here F is a holomorphic function on $\mathfrak{S}$ which up to a constant is given, on the open set where the product converges absolutely, by

$$F(\Gamma) = \prod_{\{\gamma\}} \prod_{m=0}^{\infty} (1 - q_\gamma^{1+m}),$$

where q_γ is the multiplier of $\gamma \in \Gamma$, and $\{\gamma\}$ runs over all distinct primitive conjugacy classes in $\Gamma \in \mathfrak{S}$ except the identity.

Acknowledgements S.M. was partially supported by the grant PN-II-ID-PCE 1188 265/2009.

A Simple Diffractive Boundary Value Problem on an Asymptotically Anti-de Sitter Space

Ha Pham

In this project we study the propagation of singularities (in the sense of $\mathcal{C}^\infty$ wave front set) of the solution of a model case initial – boundary value problem with glancing rays for a concave domain on an asymptotically Anti de-Sitter manifold. Singularities of solutions of linear partial differential equations are described in terms of wave front sets of distributions. In the boundaryless setting, a special case of a result by Hörmander [1] gives that for P a linear differential operator with real principal symbol p, the solution u of $Pu = 0$ has its wavefront set contained in the characteristic set $p^{-1}(0)$ and invariant under the flow generated by Hamiltonian vector field H_p. In the presence of boundary, the rays containing wavefront set which hit the boundary transversally are reflected according to Snell's law ie with energy and tangential momentum conserved (see for e.g. [2] for codimension-1 boundary and [10] for higher codimension boundary). The situation however becomes more complicated in the presence of boundary and glancing rays (i.e. tangency of the bicharacteristics). While one already has the general theorem of propagation of singularities along generalized broken bicharacteristics ([7–9] for smooth boundary, [10] for corners using a relatively permissive notion of generalized broken bicharacteristics analogous to those in the analytic case [3]), one could refine the results (i.e. make the generalized broken bicharacteristics less permissive) by considering the diffractive problem, one aspect of which studies the phenomenon of propagation of singularities into the shadow region. By results of Friedlander [4], Melrose [5], Melrose-Sjöstrand [7] etc., on smooth boundaries, there is no propagation into the shadow region and in fact, at a diffractive point, the ray carrying wavefront set does not stick to the boundary. The question remains for higher codimension settings, in particular for edges. On the other hand, asymptotically Anti de-Sitter space can be considered analogous to a reduced problem on an edge. In addition, we have the general theorem for propagation of

H. Pham (✉)
Purdue University, West Lafayette, IN, USA
e-mail: howardh@math.purdue.edu

D. Grieser et al. (eds.), *Microlocal Methods in Mathematical Physics and Global Analysis*,   133
Trends in Mathematics, DOI 10.1007/978-3-0348-0466-0_31, © Springer Basel 2013

singularities along generalized bicharachteristics by Vasy [11] on asymptotically AdS spaces for both hyperbolic and glancing cases. Hence, similar to the edge case, one would like to refine this result and study the diffractive problem for this setting. Our setting is a simple case of asympotically Anti-de Sitter spaces, which are Lorentzian manifolds modeled on Anti-de Sitter space at infinity. The metric for the problem is specified as :

$$\text{on } \mathbb{R}_x^+ \times \mathbb{R}_y^n : g = \frac{-dx^2 + (1+x)^{-1}dy_n^2 - \sum_{j=1}^{n-1} dy_j^2}{x^2}.$$

For convenience, we work with a modification of the Klein-Gordon operator by a first order derivative

$$P := \Box_g + \frac{x}{2(x+1)}x\partial_x + \lambda.$$

This modification does not change the problem in an essential way in the sense that this does not affect the propagation of singularities result on asymptotically AdS space since the principal symbol remains unchanged thus so do the bicharacteristics, and the 0-normal operator is not affected either. The diffractive condition is satisfied with $H_{\hat{p}}^2 x > 0$ where $\hat{p}$ is the principal symbol for conformal operator $\hat{P}$ ie $\hat{p} = -\xi^2 + [(1+x)\theta_n^2 - |\theta'|^2]$.

The approach adopted is motivated from the work done for a conformally related diffractive model problem by Friedlander [4] in which an explicit solution was constructed using the Airy function. The result was later greatly generalized by Melrose using a parametrix construction in [5, 6]. After taking the Fourier transform in y and denote by θ the dual variable of y, our goal is to construct a polyhomongeous conormal solution modulo a smooth function in $\dot{C}^\infty(\mathbb{R}_+^{n+1})$ which satisfies the boundary condition at $x = 0$ i.e.

$$\begin{cases} \hat{L}\hat{u} \in \dot{C}^\infty(\mathbb{R}_+^{n+1}) \\ x^{-s_-}\hat{u}\,|_{x=0} = 1 \qquad s_\pm(\lambda) = \frac{n}{2} \pm \sqrt{\frac{n^2}{4} - \lambda} \\ \hat{u} \in \exp(i\phi_{in})\mathcal{L}_{ph}(C) \quad \phi_{in} = -\frac{2}{3}\theta_n^{-1}\big[(1+x-|\hat{\theta}'|^2)^{3/2} - (1-|\hat{\theta}'|^2)^{3/2}\big]\text{sgn}\theta_n \end{cases}$$

where $\mathcal{L}_{ph}(C)$ is the set of polyhomogeneous conormal functions on some blown-up space C of $\overline{\mathbb{R}}_\theta^n \times [0, 1)_x$, and $s_\pm(\lambda)$ come from the indicial roots of the Klein Gordon operator and prescribe the asymtotic behavior of its solution on asymptotically AdS [11]. The chosen oscillatory behavior is modeled from that possessed by the specific solution Friedlander worked with in [4]. Following the same change of variable as in [4], the problem is reduced to studying the following semiclassical ODE, which at one end is a b-operator while having a scattering behavior at infinity,

$$Q = h^2(z\partial_z)^2 + h^2\left(\lambda - \frac{n^2}{4}\right) + z^3 + z^2.$$

We use different techniques near 0 and infinity to analyze the local problems: near infinity we use local resolvent bounds given by Vasy and Zworski [13] and near zero we build a local semiclassical parametrix. The next technical difficulties arise from the fact that the operator $\mathbf{Q}$ is not semiclassical elliptic, which prevents us from simply patching the two above local resolvents by partition of unity, which otherwise results in an error that is not trivial semiclassically. To overcome this problem, we adopt the method of Datchev and Vasy [12] to get a global approximate resolvent with an $O(h^\infty)$ error.

The remaining technical details of the problem involve constructing a local resolvent near 0 for $\mathbf{Q}$. For this we first need to construct a certain blown-up space of $[0,1)_z \times [0,1)_{z'} \times [0,1)_h$. The usual method (see for e.g. [14]) involves resolving the b-singularity at the corner of $\{z = z' = 0\}$ and the semiclassical singularity. In our case, we incur non-uniformity in the behavior of the normal operator on the semiclassical front face as we approach the b front face (i.e. as $z' \to 0$). This gives motivation to do a blow-up at $Z_h \to \infty$ and $z' \to 0$ where $Z = \frac{z}{z'}$ the variable associated with the b-blowup and $Z_h = \frac{Z-1}{h}$ the variable associated the blow-up along the lifted diagonal in the semiclassical face. For matter of convenience, the blown-up space we will work on results from the same idea of singularity resolution however with different order of blowing up, namely after the b-blowup we blow up the intersection of the b-front face with semiclassical face, then blow up the intersection of the lifted diagonal and the semiclassical face. There will be two additional blow-ups to desingularize the flow and to create a transition region from the scattering behavior to the boundary behavior prescribed by indicial roots. We then construct a polyhomongeous cornomal function U on this blown-up space so that

$$\mathbf{Q}U - \mathrm{SK}_{\mathrm{Id}} \in h^\infty z^\infty (z')^{\sqrt{\frac{n^2}{4}-\lambda}-\tilde{\gamma}_{\mathcal{B}_1}+2} C^\infty$$

$\mathrm{SK}_{\mathrm{Id}}$ schwartz kernel of Id, and for some constant $\tilde{\gamma}_{\mathcal{B}_1}$.

After removing singularities at the lifted diagonal, the remaining difficulty involves choosing the order in which one will remove singularities of the error to achieve the above effect.

Once constructed, a parametrix provides among other things information on the properties of the fundamental solution, which enables one to make very precise statements on the propagation of singularities, in particular in the 'shadow regions'.

Acknowledgements I would like to thank my advisor András Vasy for all his help with this project.

References

1. L. Hörmander. *On the existence and regularity of solutions of linear pseudo-differential equations*. L'Enseignement Math. **17** (1971), 99–163.
2. L. Hörmander. The analysis of Linear Partial Differential Operators, vol 4 . Springer - Verlag, 1983. 416–430.

3. G. Lebeau. *Propagation des ondes dans les variétés à coins.* Anns. Scient. Éc. Norm. Sup., 30:429–497, 19997
4. F.G. Friedlander. *The wavefront set of the solution of a simple initial-boundary value problem with glancing rays.* Math. Proc. Camb. Phil. **79** (1976), 145–159.
5. R. B. Melrose. *Microlocal parametrices for diffrative boundary value problems.* Duke Mathematical Journal. **42–4** (1975), 605–635.
6. R. B. Melrose. *Local Fourier-Airy Integral Operators.* Duke Mathematical Journal. **42–4** (1975), 583–604.
7. R. B. Melrose and J. Sjöstrand. *Singularities of boundary value problems I* Comm. Pure Appl. Math, 31:593–67, 1978.
8. R. B. Melrose and J. Sjöstrand. *Singularities of boundary value problems II* Comm. Pure Appl. Math, 35:129–168, 1982.
9. M. Taylor. *Grazing rays and reflection of singularities of solutions to wave equations.* Comm. Pure. Appl. Math., 29:1–38, 1976.
10. A. Vasy. *Propagation of singularities for the wave equation on manifold with corners.* Annals of Mathematics **168** (2008) 749–812,
11. A. Vasy. *The wave equation on asymptotically anti-de Sitter spaces* To appear in Analysis and PDE.
12. K. Datchev and A. Vasy. *Gluing semiclassical resolvent estimates via propagation of singularities* (2010) preprint.
13. A. Vasy and M. Zworski. *Semiclassical estimates in asymptotically Euclidean scattering* Comm. Math. Phys. 212(1) : 205–217, 2007.
14. R.B. Melrose, S.B. Antônio and A. Vasy. *Analytic continuation and semiclassical resolvent estimates on asymptotically hyperbolic spaces* (2010) preprint.

Quantization in a Magnetic Field

Radu Purice, Viorel Iftimie, and Marius Măntoiu

1 Introduction

Together with Marius Măntoiu we have considered quantum hamiltonians with magnetic fields and replaced the usual translations with magnetic translations, generalizing some former results from constant magnetic fields to bounded smooth magnetic fields. This approach allowed us to obtain a pseudodifferential Weyl calculus, twisted by a 2-cocycle associated to the flux of the magnetic field and we developped this calculus in colaboration with V. Iftimie. An interesting fact that we pointed out is that the algebra of observables is defined only in terms of the magnetic field without the need of a vector potential. Using these techniques we proved a number of spectral results for quantum Hamiltonians in magnetic fields. At the classical level the magnetic field may also be described by a deformation of the canonical symplectic form of the phase space and hence, a deformation of the Poisson bracket of the classical observables. At the quantum level, we can define a twisted Moyal algebra, with the Moyal product twisted by a 2-cocycle associated to the flux of the magnetic field. These two descriptions may be put together in a strict deformation quantization in the sense of M. Rieffel.

Some notations. The configuration space: $\mathcal{X} := \mathbb{R}^n$. The phase space: $\Xi := \mathbb{T}^*\mathcal{X} \cong \mathcal{X} \times \mathcal{X}'$ with momentum space $\mathcal{X}'$, the dual of $\mathcal{X}$, (canonically isomorphic to $\mathbb{R}^n$). The canonical symplectic form on Ξ:

$$\sigma\big((x, \xi), (y, \eta)\big) := < \xi, y > - < \eta, x >$$

with $< ., . >$ the duality application $\mathcal{X}' \times \mathcal{X} \to \mathbb{R}$.

R. Purice (✉) · V. Iftimie · M. Măntoiu
Institute of Mathematics of the Romanian Academy, Bucharest, Romania
e-mail: Radu.Purice@imar.ro

D. Grieser et al. (eds.), *Microlocal Methods in Mathematical Physics and Global Analysis*, 137
Trends in Mathematics, DOI 10.1007/978-3-0348-0466-0_32, © Springer Basel 2013

The magnetic field is described by a closed 2-form B on $\mathcal{X}$: $B = \sum\limits_{j,k=1}^{n}$ $B_{jk}(x)dx_j \wedge dx_k$, $B_{jk}(x) = -B_{kj}(x)$, $dB = 0$. On $\mathbb{R}^n$ there always exists a 1-form, the vector potential A such that $B = dA$. The association of a vector potential to a magnetic field B is highly non unique and we have that $A - A' = \nabla\Phi \Leftrightarrow dA = dA' = B$. If B has components of class $C_{pol}^{\infty}(\mathcal{X})$, then one can always find a vector potential with components of class $C_{pol}^{\infty}(\mathcal{X})$.

2 A 'Magnetic' Weyl System

The idea we propose is to use the unitary groups associated to the $2n$ self-adjoint operators $\{Q_1, \ldots, Q_n\}$, $\{\Pi_1^A, \ldots, \Pi_n^A\}$ defined previously and their real linear combinations, and define the Magnetic Weyl system (Just use the Kato-Trotter formula):

$$W^A((x,\xi)) := e^{-i<\xi,(Q+x/2)>}\, e^{-i\int_{[Q,Q+x]}A}\, e^{i<x,P>}.$$

- For any test function $f : \Xi \to \mathbb{C}$ we define the associated magnetic Weyl operator:

$$\mathfrak{Op}^A(f) := \int_\Xi dX\,\hat{f}(X)W^A(X) \in \mathbb{B}[\mathcal{H}]$$

- In fact for any tempered distribution $F \in \mathcal{S}'(\Xi)$ we can define the linear operator:

$$\mathfrak{Op}^A(F) := \int_\Xi dX\,\hat{F}(X)W^A(X) \in \mathbb{B}[\mathcal{S}(\mathcal{X}); \mathcal{S}'(\mathcal{X})]$$

- It defines a linear bijection [11].
- The Schrödinger representations associated to any two gauge-equivqlent vector potentials are unitarily equivqlent (gauge covariance): $A' = A + d\varphi \Rightarrow \mathfrak{Op}^{A'}(f) = e^{i\varphi(Q)}\mathfrak{Op}^A(f)e^{-i\varphi(Q)}$.

3 A 'Magnetic' Moyal Algebra

The above 'magnetic' functional calculus induces a *magnetic composition* on the complex linear space of test functions: $\mathfrak{Op}^A(f\,\sharp^B g) := \mathfrak{Op}^A(f) \cdot \mathfrak{Op}^A(g)$. It only depends on the magnetic field B! in fact on the 'magnetic' deformation of the symplectic form on Ξ.

Theorem 1 ([11]). *For a magnetic field B with components of class $C_{\mathrm{pol}}^{\infty}(\mathcal{X})$, the composition $\sharp^{B}$ defines a bilinear map*

$$\mathcal{S}(\Xi) \times \mathcal{S}(\Xi) \ni (\phi, \psi) \mapsto \phi\sharp^{B}\psi \in \mathcal{S}(\Xi)$$

that is jointly continuous (for the usual Fréchet topology on $\mathcal{S}(\Xi)$).

Proposition 1 ([11]). *For a magnetic field B with components of class $C_{\mathrm{pol}}^{\infty}(\mathcal{X})$, we have:*

$$\int_{\Xi}(\phi\sharp^{B}\psi)(X)\,dX \;=\; \int_{\Xi}\phi(X)\,\psi(X)\,dX, \qquad \forall(\phi, \psi) \in \left(\mathcal{S}(\Xi)\right)^{2},$$

$$\int_{\Xi}(\phi\sharp^{B}\psi)(X)\chi(X)dX \;=\; \int_{\Xi}\phi(X)\big(\psi\sharp^{B}\chi\big)(X)dX, \quad \forall(\phi, \psi, \chi) \in \left(\mathcal{S}(\Xi)\right)^{3}.$$

We ca extend the product $\sharp^{B}$ by duality to bilinear maps:

$$\mathcal{S}'(\Xi)\sharp^{B}\mathcal{S}(\Xi) \to \mathcal{S}'(\Xi); \quad \mathcal{S}(\Xi)\sharp^{B}\mathcal{S}'(\Xi) \to \mathcal{S}'(\Xi).$$

The magnetic Moyal algebra. We set:

$$\mathfrak{M}^{B}(\Xi) := \big\{F \in \mathcal{S}'(\Xi) \mid F\sharp^{B}\phi \in \mathcal{S}(\Xi), \phi\sharp^{B}F \in \mathcal{S}(\Xi), \forall \phi \in \mathcal{S}(\Xi)\big\}$$

This defines a *-algebra for the *composition* $\sharp^{B}$ and the usual complex conjugation as *-*conjugation*.

Proposition 2 ([11]). *The space of indefinitely differentiable functions with uniform polynimial growth on Ξ is contained in $\mathfrak{M}^{B}(\Xi)$.*

Proposition 3 ([5]). *If the magnetic field B has components of class $C_{\mathrm{pol}}^{\infty}(\mathcal{X})$, for $m \in \mathbb{R}$ and $0 \le \delta \le \rho \le 1$ we have $S_{\rho,\delta}^{m}(\Xi) \subset \mathfrak{M}^{B}(\Xi)$.*

The norm. The family: $\mathfrak{C}^{B}(\Xi) := \big\{F \in \mathcal{S}'(\Xi) \mid \mathfrak{Op}^{A}(F) \in \mathbb{B}[L^{2}(\mathcal{X})]\big\}$ does only depend on the magnetic field B.

On $\mathfrak{C}^{B}(\Xi)$ we can define the map: $\|F\|_{B} := \|\mathfrak{Op}^{A}(F)\|_{\mathbb{B}[L^{2}(\mathcal{X})]}$ that does not depend on the choice of A and is in fact a C*-norm on $\mathfrak{C}^{B}(\Xi)$. $\mathfrak{C}^{B}(\Xi)$ is a C*-algebra isomorphic to $\mathbb{B}[L^{2}(\mathcal{X})]$.

Theorem 2 ([5]). *If the magnetic field B has components of class $BC^{\infty}(\mathcal{X})$, then $S_{\rho,\rho}^{0}(\Xi)$, with $0 \le \rho < 1$ and $S_{\rho,\delta}^{0}(\Xi)$, with $0 \le \delta < \rho \le 1$ are contained in $\mathfrak{C}^{B}(\Xi)$ and there exist two constants $c(n) \in \mathbb{R}_{+}$ and $p(n) \in \mathbb{N}$, depending only on the dimension n of the space $\mathcal{X}$, such that we have the estimation (where $|F|_{(p,q)}$ are the seminorms defining the topology of $S_{\rho,\delta}^{0}(\Xi)$):*

$$\|F\|_{B} \le c(n)|F|_{(p(n),p(n))}.$$

4 Some Operator-Algebraic Aspects

Let us first observe that with any Weyl system associated to a choice of a magnetic potential A, we have:

- A representation of the C^*-algebra $BC(\mathcal{X})$ (considered as functions on Ξ constant along $\mathcal{X}'$)

$$BC(\mathcal{X}) \ni f \mapsto \mathfrak{Op}^A(f) \equiv f(Q) \in \mathbb{B}[\mathcal{H}], \quad [f(Q)u](x) = f(x)u(x)$$

(by usual functional calculus associated to V.

- A *twisted* unitary representation of $\mathcal{X}$

$$\mathcal{X} \ni x \mapsto \mathfrak{Op}^A(\mathfrak{e}_x) \equiv U^A(x) \in \mathbb{B}[\mathcal{H}], \text{ where: } \mathfrak{e}_x(y, \eta) := e^{ix \cdot \eta}$$

In fact one has $U^A(x)U^A(y) = \omega^B(Q; x, y)U^A(x + y)$,
where: $\omega^B(q; x, y) = \exp\left\{-i \int_{\mathcal{T}(q, q+x, q+x+y)} B\right\}$ with $\mathcal{T}(q_1, q_2, q_3)$ the triangle of vertices q_1, q_2, q_3.

We have:
$$U^A(x)f(Q)[U^A(x)]^{-1} = f(Q + x) = [\tau_x f](Q).$$

We put thus into evidence the following *twisted C^*-dynamical system* $\{\mathcal{X}, BC_u(\mathcal{X}), \tau, \omega^B\}$. Given a twisted C^*-dynamical system $\{\mathcal{X}, \mathcal{A}, \tau, \omega\}$ let us consider:

- The linear space: $L^1(\mathcal{X}; \mathcal{A})$ of measurable functions $f : \mathcal{X} \to \mathcal{A}$ such that the positive function $\mathcal{X} \ni x \mapsto \|f(x)\|_{\mathcal{A}}$ is integrable with respect to the Haar measure $\mu_{\mathcal{X}}$ on $\mathcal{X}$
- The (τ, ω) crossed-product:

$$(f \bowtie_\tau^\omega g)(x) := \int_{\mathcal{X}} dy\ \tau_{\frac{y-x}{2}}[f(y)] \cdot \tau_{\frac{y}{2}}[g(x - y)] \cdot \tau_{-\frac{x}{2}}[\omega(y, x - y)]$$

with $\cdot$ the product in the algebra $\mathcal{A}$.
- The involution $f^{\bowtie}(x) := \tau_{-\frac{x}{2}}[\omega(x, -x)^{-1}]\overline{f(-x)}$
- The C^*-enveloping norm $\|.\|_*$

Its closure with respect to $\|.\|_*$ is **the twisted crossed-product C^*-algebra** $\mathcal{A} \bowtie_\tau^\omega \mathcal{X}$.

It is a standard matter that all the covariant representations of a twisted C^*-dynamical system $\{\mathcal{X}, \mathcal{A}, \rho, \omega\}$ are in a one-to-one correspondence with the representations of the twisted crossed-product $\mathcal{A} \bowtie_\rho^\omega \mathcal{X}$.

Some computations allow us to prove that the partial Fourier transformation:

$$\mathcal{F} : L^1(\mathcal{X}; BC_u(\mathcal{X})) \to BC(\Xi), \quad [\mathcal{F}f](x, \xi) := \left[\int_{\mathcal{X}} dy e^{-i<\xi, y>} f(y)\right](x)$$

gives an isometric embedding: $\mathcal{F} : BC_u(\mathcal{X}) \bowtie_\tau^{\omega^B} \mathcal{X} \hookrightarrow \mathfrak{C}^B$.

The important achievement of the above formalism is that it provides us with a procedure to associate to any translation invariant sub-C^*-algebra $\mathcal{A}$ of $BC_u(\mathcal{X})$, a sub-C^*-algebra $\mathfrak{C}_{\mathcal{A}}^B$ of $\mathfrak{C}^B$:

$$\mathfrak{C}_{\mathcal{A}}^B(\Xi) := \mathcal{A}(\mathcal{X}) \rtimes_{\tau}^{\omega^B} \mathcal{X}.$$

Proposition 4 ([11]). *For any magnetic field B with components of class $BC^\infty(\mathcal{X})$ we have*

$$C_0(\mathcal{X}) \rtimes_{\tau}^{\omega^B} \mathcal{X} \cong \mathbb{B}_\infty[\mathcal{H}]$$

(the ideal of compact operators on $\mathcal{H}$).

5 A 'Magnetic' Pseudo-differential Calculus

Definition 1. Choosing any vector potential A for B we define the associated classes of *magnetic* pseudodifferential operators on $\mathcal{H} := L^2(\mathcal{X})$ with Hörmander type symbols:

$$\Psi_{\rho,\delta}^m(A) := \mathfrak{Op}^A[S_{\rho,\delta}^m(\Xi)].$$

Proposition 5. *For $0 \le \delta < \rho \le 1$ or $0 \le \delta = \rho < 1$ we have that*

$$\Psi_{\rho,\delta}^0(A) \subset \mathfrak{C}^B(\Xi).$$

Theorem 3 ([5]). *If the magnetic field B has components of class $C_{\mathrm{pol}}^\infty(\mathcal{X})$, for any m_1 and m_2 in $\mathbb{R}$ and for any $0 \le \delta \le \rho \le 1$ we have:*

$$S_{\rho,\delta}^{m_1}(\Xi) \, \sharp^B \, S_{\rho,\delta}^{m_2}(\Xi) \subset S_{\rho,\delta}^{m_1+m_2}(\Xi).$$

Definition 2. Suppose that the magnetic field B has components of class $BC^\infty(\mathcal{X})$ and suppose chosen a vector potential A for it. For any $m > 0$ we define the complex linear space:

$$\mathcal{H}_A^m(\mathcal{X}) := \left\{ u \in L^2(\mathcal{X}) \mid \mathfrak{p}_m^A u \in L^2(\mathcal{X}) \right\}$$

with

$$\mathfrak{p}_m^A := \mathfrak{Op}^A(\wp_m), \quad \wp_m(x,\xi) := <\xi>^m \equiv (1 + |\xi|^2)^{m/2}, \quad \forall m > 0.$$

Definition 3. Suppose that the magnetic field B has components of class $BC^\infty(\mathcal{X})$ and suppose chosen a vector potential A. For any $m > 0$ we define the space $\mathcal{H}_A^{-m}(\mathcal{X})$ as the dual space of $\mathcal{H}_A^m(\mathcal{X})$ with the dual norm:

$$\|\phi\|_{(-m,A)} := \sup_{u \in \mathcal{H}_A^m(\mathcal{X}) \setminus \{0\}} \frac{|<\phi, u>|}{\|u\|_{(m,A)}}$$

that induces a scalar product.

We also denote $\mathcal{H}_A^0(\mathcal{X}) := L^2(\mathcal{X})$.

Definition 4. For $m > 0$ a symbol $F \in S_{\rho,\delta}^m(\Xi)$ is said to be **elliptic** if there exist two positive constants R and C such that for any $(x, \xi) \in \Xi$ with $|\xi| \geq R$ one has that $|F(x, \xi)| \geq C < \xi >^m$.

Theorem 4 ([5]).

- *Suppose that the magnetic field B has components of class $BC^\infty(\mathcal{X})$ and suppose chosen a vector potential A for it.*
- *Suppose $m \geq 0$ and $F \in S_{\rho,\delta}^m(\Xi)$ is a real symbol (elliptic if $m > 0$), with either $0 \leq \delta < \rho \leq 1$ or $\delta = \rho \in [0, 1)$.*
- *Then the operator*

$$\mathfrak{Op}^A(F) : \mathcal{H}_A^m(\mathcal{X}) \to L^2(\mathcal{X})$$

  is self-adjoint.
- *If $F \geq 0$ then $\mathfrak{Op}^A(F)$ is lower semibounded and we have a strong Gårding inequality.*
- *If A is chosen in $C_{\text{pol}}^\infty(\mathcal{X})$, then $\mathfrak{Op}^A(F)$ is essentially self-adjoint on $\mathcal{S}(\mathcal{X})$.*

6   Some Results for Quantum Hamiltonians

6.1   Anisotropic Magnetic Hamiltonians

We are interested in anisotropic problems [8,9]. The anisotropy will be characterized by a C^*-subalgebra algebra:

$$\mathcal{A}(\mathcal{X}) \subset BC_u(\mathcal{X})$$

satisfying:

- $\mathcal{A}(\mathcal{X})$ is a unital C^*-subalgebra of $BC_u(\mathcal{X})$
- $\mathcal{A}(\mathcal{X})$ is left invariant by all the translations τ_x (with $x \in \mathcal{X}$).
- $C_0(\mathcal{X}) \subset \mathcal{A}(\mathcal{X})$.

Then

- $$\left[\mathcal{A}(\mathcal{X}) \rtimes_\tau^{\omega^B} \mathcal{X}\right] \Big/ \left[C_0(\mathcal{X}) \rtimes_\tau^{\omega^B} \mathcal{X}\right] \cong \left[\mathcal{A}(\mathcal{X}) \big/ C_0(\mathcal{X})\right] \rtimes_\tau^{\omega^B} \mathcal{X}$$

and we have a very effective procedure to obtain *a decomposition of the essential spectrum as closure of the union of the spectra of the 'asymptotic limits' of the Hamiltonian.*

6.2   Continuity of the Spectra

Consider a family of Hamiltonians $\{h^\epsilon\}_{\epsilon \in I}$ with $I \subset \mathbb{R}$ a compact interval, such that

- $h^\epsilon \in S_1^m(\Xi)$ elliptic with $m > 0$, for each $\epsilon \in I$,
- The map $I \ni \epsilon \mapsto h^\epsilon \in S_1^m(\Xi)$ is continuous for the Fréchet topology on $S^m(\Xi)$;
- There exist $C \in \mathbb{R}_+$ such that $h^\epsilon \geq -C$, $\forall \epsilon \in I$.

We are given a family of magnetic fields $\{B^\epsilon\}_{\epsilon \in I}$ with the components $B_{jk}^\epsilon \in BC^\infty(\mathcal{X})$ such that the map $I \ni \epsilon \mapsto B_{jk}^\epsilon \in BC^\infty(\mathcal{X})$ is continuous for the Fréchet topology on $BC^\infty(\mathcal{X})$.

Definition 5. Let I be a compact interval and suppose given a family $\{\sigma^\epsilon\}_{\epsilon \in I}$ of closed subsets of $\mathbb{R}$.

1. The family $\{\sigma^\epsilon\}_{\epsilon \in I}$ is called *outer continuous* at $\epsilon_0 \in I$ if for any compact $K \subset \mathbb{R}$ such that $K \cap \sigma^{\epsilon_0} = \emptyset$, there exists a neighborhood $V_K^{\epsilon_0}$ of ϵ_0 with $K \cap \sigma^\epsilon = \emptyset$, $\forall \epsilon \in V_K^{\epsilon_0}$.
2. The family $\{\sigma^\epsilon\}_{\epsilon \in I}$ is called *inner continuous at* $\epsilon_0 \in I$ if for any open $\mathcal{O} \subset \mathbb{R}$ such that $\mathcal{O} \cap \sigma^{\epsilon_0} \neq \emptyset$, there exists a neighborhood $V_\mathcal{O}^{\epsilon_0} \subset I$ of ϵ_0 with $\mathcal{O} \cap \sigma^\epsilon \neq \emptyset$, $\forall \epsilon \in V_\mathcal{O}^{\epsilon_0}$.
3. The family $\{\sigma^\epsilon\}_{\epsilon \in I}$ is called *continuous at* $\epsilon_0 \in I$ if it is both inner and outer continuous.

Theorem 5 ([1]). *Suppose given a compact interval $I \subset \mathbb{R}$, a family of classical Hamiltonians $\{h^\epsilon\}_{\epsilon \in I}$ and a family of magnetic fields $\{B^\epsilon\}_{\epsilon \in I}$ satisfying the above hypothesis. Let us consider the family of quantum Hamiltonians $H^\epsilon := \mathfrak{Op}^{A^\epsilon}(h^\epsilon)$ for some choice of a vector potential A^ϵ for B^ϵ. Then the spectra $\sigma^\epsilon := \sigma(H^\epsilon) \subset \mathbb{R}$ form a continuous family of subsets at any point $\epsilon \in I$.*

6.3 Eigenfunction Decay

Theorem 6 ([3]).
Let us suppose that

- $h \in S_1^m(\mathbb{R}^d)$ *(with $m > 0$) is elliptic;*
- *The magnetic field B has components of class $BC^\infty(\mathcal{X})$;*
- *We fixed a vector potential A for the magnetic field having components of class $BC^\infty(\mathcal{X})$.*

Let $\lambda \in \sigma_{disc}(\mathfrak{Op}^A(h))$ and $u \in \mathrm{Ker}(\mathfrak{Op}^A(h) - \lambda)$. Then $u \in \mathcal{S}(\mathcal{X})$.

References

1. N. Athmouni, M. Măntoiu, and R. Purice: On the continuity of spectra for families of magnetic pseudodifferential operators. Journal of Mathematical Physics **51**, 083517 (2010); doi:10.1063/1.3470118 (15 pages).

2. B. Helffer and R. Purice: Magnetic calculus and semiclassical trace formulas. Journal of Physics A: Mathematical and Theoretical **43** (2010) 474028 (21pp)
3. V. Iftimie, R. Purice: Eigenfunctions decay for magnetic pseudodifferential operators, Journal of Mathematical Physics, **52** (9) (2011), doi:10.1063/ 1.3642622 (11 pages).
4. V. Iftimie, M. Măntoiu, R. Purice: Commutator Criteria for Magnetic Pseudodifferential Operators. Comm. Partial Diff. Eq. **35** (2010), 1058?1094.
5. V. Iftimie; M. Măntoiu; R. Purice: Magnetic pseudodifferential operators, Publications of RIMS, **43** (2007), no. 3, 585?623.
6. M.V. Karasev and T.A. Osborn, Symplectic Areas, Quantization and Dynamics in Electromagnetic Fields, J. Math. Phys. **43** (2002), 756–788.
7. M.V. Karasev and T.A. Osborn, Quantum Magnetic Algebra and Magnetic Curvature, J. Phys.A **37** (2004), 2345–2363.
8. M. Lein, M. Măntoiu and S. Richard: *Magnetic Pseudodifferential Operators with Coefficients in C*-algebras*, Publ. Res. Inst. Math. Sci. **46** no. 4, 755-788, (2010).
9. M. Măntoiu; R. Purice; S. Richard: Spectral and propagation results for magnetic Schodinger operators; a C*-Algebraic framework, Journal of Functional Analysis, **250** (2007), 42–67;
10. M. Măntoiu; R. Purice: Radu Strict deformation quantization for a particle in a magnetic field. J. Math. Phys. **46** (2005), no. 5, 052105, 15 pp.
11. M. Măntoiu; R. Purice: The magnetic Weyl calculus. J. Math. Phys. **45** (2004), no. 4, 1394–1417.

Price's Law on Black Hole Space-Times

Daniel Tataru

The aim of the talk was to provide an overview of recent and ongoing work concerning global in time decay properties for the wave equation on asymptotically flat space-times. Parts of this work are joint with the following collaborators: Jeremy Marzuola, Jason Metcalfe and Mihai Tohaneanu. Some of this research was motivated by problems in general relativity concerning the decay properties for the wave equation on Schwarzschild and Kerr backgrounds. Partly for this reason, all the results are presented in $3 + 1$ space dimensions.

We consider decay estimates for the forward wave equation

$$(\Box_g + V)u = f, \qquad u(0) = u_0, \quad \partial_t u(0) = u_1 \tag{1}$$

For the metric g and the potential V we consider three cases:

Case A: g is a smooth Lorenzian metric in $\mathbb{R} \times \mathbb{R}^3$, with the following properties: (i) The level sets $t = const$ are space-like, and (ii) g is asymptotically flat, i.e. $g = m + o_r(1)$ and $V = o_r(r^{-2})$ as $|x| = r \to \infty$, where m stands for the Minkowski metric.

Case B: g is a smooth Lorenzian metric in an exterior domain $\mathbb{R} \times \mathbb{R}^3 \setminus B(0, R_0)$ which satisfies (i) and (ii) above, as well as (iii) the lateral boundary $\mathbb{R} \times \partial B(0, R_0)$ is outgoing space-like.

Case C: g is a smooth Lorenzian metric in an exterior domain $\mathbb{R} \times \mathbb{R}^3 \setminus B(0, R_0)$ which satisfies (i) and (ii) above, as well as (iv) the lateral boundary $\mathbb{R} \times \partial B(0, R_0)$ is time-like, and u satisfies a Dirichlet boundary condition.

The second case is modeled after the Schwarzschild and Kerr metrics, which satisfy the above conditions in suitable advanced time coordinates. There the parameter R_0 is chosen so that $0 < R_0 < 2M$ in the case of the Schwarzschild

D. Tataru (✉)
University of California, Berkeley, CA, USA
e-mail: tataru@math.berkeley.edu

D. Grieser et al. (eds.), *Microlocal Methods in Mathematical Physics and Global Analysis*, 145
Trends in Mathematics, DOI 10.1007/978-3-0348-0466-0_33, © Springer Basel 2013

metric, respectively $r^- < R_0 < r^+$ in the case of Kerr, so that the exterior of the R_0 ball contains a neighbourhood of the event horizon.

One question we ask here is what is the sharp pointwise decay rate for linear waves with, say, smooth, compactly supported data. By the finite speed of propagation, the solutions will be localized in a forward light cone; it is also natural to expect a t^{-1} decay rate along the light cone. The interesting problem is then what is the decay rate inside the cone.

In the Minkowski case Huygens principle holds, so the solution will vanish inside a smaller light cone. Adding a small compactly supported potential generates some tails inside the cone, but these decay exponentially. However, adding polynomial tails to either V or the metric it seems natural to expect only polynomial decay rates inside the cone. In addition, for large perturbations one needs to also consider trapping issues; indeed, in Schwarzschild/Kerr and perturbations thereof trapping necessarily occurs.

This work is motivated in part by some heuristic computations of a physicist, Robert Price, who derived a t^{-3} local decay rate for the case of the Schwarzschild space-time. This corresponds to a metric $g \sim m + O(1/r)$ and/or a potential decaying like $O(1/r^3)$; here m represents the Minkowski metric. The t^{-3} decay rate became known as "Price's Law". In addition to the local decay given by Price's Law, it is also of interest to understand the decay in the entire light cone. The main result can be stated as follows:

Theorem 1 ([11, 22]). *Consider either case A,B,C. Suppose that the metric g and the potential V satisfy*

$$g = m + O_{radial}(r^{-1}) + o(r^{-1}), \qquad g = m + O_{radial}(r^{-3}) + o(r^{-3})$$

If in addition some suitable local assumptions hold then we have the following pointwise decay estimates for solutions to (1) with smooth localized data

$$|u(t,x)| \le \frac{C}{\langle t + |x| \rangle \langle t - |x| \rangle^2}, \qquad |\partial_t u(t,x)| \le \frac{C}{\langle t + |x| \rangle \langle t - |x| \rangle^3} \qquad (2)$$

The suitable local assumptions above are of two types:

I. Energy estimates, where the aim is to obtain uniform in time bounds,

$$\|\nabla_{x,t} u\|_{L^\infty L^2} \le \|\nabla_{x,t} u(0)\|_{L^2} + \|f\|_{L_t^1 L_x^2}$$

II. Local energy decay, i.e. integrated energy decay in compact spatial regions:

$$\|\nabla_{x,t} u\|_{LE} + \|\langle r \rangle^{-1} u\|_{LE} \le \|\nabla_{x,t} u(0)\|_{L^2} + \|f\|_{LE^*}$$

where the dual LE and LE^* norms are defined using the partition of the space-time $\mathbb{R}^+ \times \mathbb{R}^3$ into dyadic regions $A_m = \{\langle r \rangle \approx 2^m\}$

$$\|v\|_{LE} = \sup_m \|\langle r \rangle^{-\frac{1}{2}} v\|_{L^2_{x,t}(A_m)}, \qquad \|f\|_{LE*} = \sum_m \|\langle r \rangle^{\frac{1}{2}} f\|_{L^2_{x,t}(A_m)}$$

In the case when trapping occurs the energy estimates remain unchanged, but it is natural to weaken somewhat the local energy decay norms on the trapped set. This makes no difference in the above theorem.

At this point, these two types of estimates are known to be valid for small perturbations of the Minkowski and Schwarzschild/Kerr metrics. In the Minkowski case, no time decay is needed for the perturbation. However, in the Schwarzschild/Kerr case a t^{-1-} decay is assumed near the trapped set. More generally, one can split the problem into a low frequency part, where the enemy is represented by bound states, and a high frequency part, where the issue is trapping.

References

1. Serge Alinhac. On the Morawetz–Keel-Smith-Sogge inequality for the wave equation on a curved background. *Publ. Res. Inst. Math. Sci.*, 42(3):705–720, 2006.
2. Lars Andersson and Pieter Blue. Hidden symmetries and decay for the wave equation on the Kerr spacetime. arXiv:0908.2265.
3. Pieter Blue and Jacob Sterbenz. Uniform decay of local energy and the semi-linear wave equation on Schwarzschild space. *Comm. Math. Phys.*, 268(2):481–504, 2006.
4. Mihalis Dafermos and Igor Rodnianski. Lectures on black holes and linear waves. arXiv:0811.0354.
5. Mihalis Dafermos and Igor Rodnianski. A proof of the uniform boundedness of solutions to the wave equation on slowly rotating Kerr backgrounds. arXiv:0805.4309.
6. Mihalis Dafermos and Igor Rodnianski. The red-shift effect and radiation decay on black hole spacetimes. *Comm. Pure Appl. Math.*, 62(7):859–919, 2009.
7. F. Finster, N. Kamran, J. Smoller, and S.-T. Yau. Decay of solutions of the wave equation in the Kerr geometry. *Comm. Math. Phys.*, 264(2):465–503, 2006.
8. Felix Finster and Joel Smoller. A time-independent energy estimate for outgoing scalar waves in the Kerr geometry. *J. Hyperbolic Differ. Equ.*, 5(1):221–255, 2008.
9. S. W. Hawking and G. F. R. Ellis. *The large scale structure of space-time*. Cambridge University Press, London, 1973. Cambridge Monographs on Mathematical Physics, No. 1.
10. Jeremy Marzuola, Jason Metcalfe, Daniel Tataru and Mihai Tohaneanu. Strichartz estimates on Schwarzschild black hole backgrounds. Comm. Math. Phys. 293 (2010), no. 1, 3783.
11. Jason Metcalfe, Daniel Tataru and Mihai Tohaneanu. Price's Law on Nonstationary Space-times, arXiv:1104.5437
12. Johann Kronthaler. Decay rates for spherical scalar waves in the Schwarzschild geometry. preprint, arXiv:0709.3703.
13. Jonathan Luk. Improved decay for solutions to the linear wave equation on a schwarzschild black hole. arXiv:0906.5588.
14. Jeremy Marzuola, Jason Metcalfe, and Daniel Tataru. Strichartz estimates and local smoothing estimates for asympotottically flat Schrödinger equations. *J. Funct. Anal.*, 255(6):1497–1553, 2008.
15. Jason Metcalfe and Daniel Tataru. Global parametrices and dispersive estimates for variable coefficient wave equations. Math. Ann., to appear, arXiv:0707.1191, 2007.
16. Cathleen S. Morawetz. Exponential decay of solutions of the wave equation. *Comm. Pure Appl. Math.*, 19:439–444, 1966.

17. Richard H. Price. Nonspherical perturbations of relativistic gravitational collapse. I. Scalar and gravitational perturbations. *Phys. Rev. D (3)*, 5:2419–2438, 1972.
18. Avy Soffer Roland Donninger, Wilhelm Schlag. A proof of price's law on Schwarzschild black hole manifolds for all angular momenta. arXiv:0908.4292.
19. Nikodem Szpak, Piotr Bizoń, Tadeusz Chmaj, and Andrzej Rostworowski. Linear and nonlinear tails. II. Exact decay rates and spherical symmetry. *J. Hyperbolic Differ. Equ.*, 6(1):107–125, 2009.
20. Daniel Tataru. Parametrices and dispersive estimates for Schrödinger operators with variable coefficients. *Amer. J. Math.*, 130(3):571–634, 2008.
21. Daniel Tataru and Mihai Tohaneanu. Local energy estimates on Kerr black hole backgrounds. arXiv:0810.5766.
22. Daniel Tataru. Local decay of waves on asymptotically flat stationary space-times. AJM, to appear arXiv:0910.5290
23. Mihai Tohaneanu. Strichartz estimates on Kerr black hole backgrounds. AMS Transactions, to appear